M.I.T. Report No. 19

# Design of Lossy Filters

J. W. Craig

The M.I.T. Press
Cambridge, Massachusetts, and London, England

Originally published as Technical Report 411 *Theory and Design Data for Uniformly Dissipative, Doubly Terminated Bandpass and Lowpass Filters* in 1966 by the Lincoln Laboratory.

ISBN 0 262 03038 1

Library of Congress catalog card number: 70-107988.

Printed in the United States of America.

CONTENTS

## PREFACE

The material presented in this book is intended primarily to be used for the quick and accurate design and realization of narrow bandpass filters whose tuned circuits have uniform parasitic dissipation - a condition which closely approximates the usual practical situation. Lowpass and wide bandpass filters can also be designed using the data, although in some situations other methods of accounting for losses in these circuits are preferable. To use the material, one does not need to be a filter or circuit theory specialist. Anyone with a basic knowledge of electric circuits and components and some practical experience in lay-out and construction should be able to use this data to design and construct filters with close to the predicted characteristics.

The work reported here was done while the author was a Staff Member at MIT Lincoln Laboratory and is essentially identical to that in their Technical Report No. 411, 24 February 1966, which has received limited circulation.

The author is grateful to the Lincoln Laboratory Editorial Committee for permission to publish this material, in particular to Dr. Joseph Mindel and Mr. Paul Rosen for their support. Mr. Paul R. Drouilhet, Jr., originally suggested the problem and suggested improvements in the original manuscript. Mrs. Nancy Rawson wrote the programs for generating the data and solved the set of equations for $n = 5$, and Mrs. Ann Carpenter plotted and drew most of the curves. It is a pleasure to thank them for their contributions.

John W. Craig
Sperry Rand Research Center
Sudbury, Massachusetts
February 1970

## I. INTRODUCTION

In the design of frequency selective filters, it is often necessary to take into account the parasitic dissipation of the reactive components which will be used to realize the filter. In this report, we present design parameters for filters having Butterworth, Chebyshev, and Bessel transfer functions for a specific, but quite useful, distribution of parasitic dissipation. The particular case considered is that in which all circuit elements are assumed to have the same value of unloaded Q (such a circuit is said to be uniformly lossy). The resulting designs are most useful for the design of narrow bandpass filters realized as a cascade of coupled resonant circuits; here the uniformly lossy assumption reduces to that of equal Q for each of the coupled resonant circuits. However, the design parameters apply equally well to the realization of lowpass and wide bandpass filters. In all cases the design is based on a normalized, equivalent, lowpass filter.

For each of the filter characteristics, design parameters are given for filters having two through five poles. In the case of Chebyshev filters, pass-band ripples of 0.001, 0.01, 0.03, 0.10, 0.30, and 1.0 dB are included.

In addition to the design data, normalized gain curves are given from which one can calculate the magnitude of the transfer function of each filter at the center frequency. Useful characteristics of the normalized, lowpass filters are also presented, including plots of unit impulse and unit step response and attenuation, phase, and group delay. From the impulse and step responses, the envelope of the corresponding responses of a narrow bandpass filter can be determined.

The data available previously for the design of bandpass or lowpass filters fall into two categories:

(a) Lossless design with resistive source and load or with a resistive termination at one end only, and

(b) Uniformly lossy design with a resistive termination at one end only.

Such data can be found in Refs. 1 through 11. To the author's knowledge, these are the first published data for the design of dissipative filters with resistive source and load.

The remainder of the report is organized as follows. In Section II, the normalizations used for the Butterworth, Chebyshev and Bessel lowpass transfer functions are explained (these correspond to wide current usage), and tables of numerical values are given for the poles and denominator polynomial coefficients of the transfer functions. In addition, general expressions are given (as a function of the passband ripple) for the polynomial coefficients of Chebyshev transfer functions. These can be used in the design equations given in Appendix C when it is necessary to design a Chebyshev filter with a ripple which is not equal to one of the values given above. The frequency and time responses are also presented graphically in Section II. Section III is a guide to the design data and their application. Design parameters are defined and related to the circuit element values for specific filter realizations. Frequency transformations that convert a bandpass or a lowpass filter into an equivalent normalized lowpass filter and gain normalizations are discussed. Seven figures illustrate explicitly the definitions of the normalized design parameters for the various network configurations of interest. Use of the design data to realize a specific filter is illustrated by means of an example. The design data are presented in Section IV.

In Appendix A filter alignment techniques based on the work of Dishal[20] are discussed, and in Appendix B the various circuits used to realize the filters are analyzed and their transfer functions are derived. Many of these results were first given by Dishal[12] in 1949, but he did not publish the derivation which is carried out here for the sake of completeness. A few intermediate results are believed to be new, although the final results are not. The design parameters arise naturally in the course of the work.

Appendix C presents solutions for the design parameters in terms of the denominator polynomial coefficients of the desired transfer function. The methods used to solve the design problem are those of Dishal[12] and Wagner,[13] but the solution for a five-pole transfer function apparently is new. These solutions are the basis for the design curves presented in this report and can also be used for the design of filters with other all-pole transfer functions. To realize a given transfer function, it is possible to have more than one set of values of the design parameters. An alternate viewpoint for these multiple solutions is obtained from modern network synthesis, and some previously unobserved solutions for Bessel transfer functions are pointed out.

Numerous checks of the data given here against those in Refs. 1, 2, and 5 have been made, and no disagreement has been found. This checks, of course, only those areas in which there is overlap between the data presented here and those given in the references. As further verification, the data were spot-checked at many points to see that the required transfer function resulted. Also, many filters have been designed successfully using these data.

## II. NORMALIZED TRANSFER FUNCTIONS

In this section we define the normalized, lowpass Butterworth, Chebyshev, and Bessel transfer functions that are the basis of the design data, and we explain the bandwidth factor w which is used in the transformation of the actual frequency $\omega = 2\pi f$ to the normalized frequency $\Omega$. For the convenience of the designer, curves of attenuation, phase, and group delay as well as unit impulse and unit step response are given for these transfer functions.

### A. TRANSFER FUNCTIONS

As is conventional for Butterworth filters, the lowpass transfer function has a half-power frequency of 1 rad/sec, and the squared magnitude of the transfer function is

$$|t_{BU}(i\Omega)|^2 = \frac{1}{1 + \Omega^{2n}} \quad . \tag{2-1}$$

The parameter n, the number of poles, specifies a particular Butterworth transfer function which we will designate by the notation BU n. Thus, BU 4 refers to a four-pole Butterworth transfer function. This notation is an adaptation of that used by Saal[6] for Cauer parameter filters. The transfer function is

$$t_{BU}(\lambda) = \frac{1}{Q_n(\lambda)} \tag{2-2}$$

where $Q_n(\lambda) = \lambda^n + q_{n-1}\lambda^{n-1} + \ldots + q_1\lambda + q_0$, and $q_0 = 1$ for this case. The zeros and coefficients of $Q_n(\lambda)$ are given in Tables I and II.

For Chebyshev characteristics the lowpass transfer function has equal-ripple behavior out to 1 rad/sec, and the squared magnitude of its transfer function is

$$|t_{CH}(i\Omega)|^2 = \frac{1}{1 + \epsilon^2 \, T_n^2(\Omega)} \quad . \tag{2-3}$$

Here $T_n(\Omega)$ is the Chebyshev polynomial of the first kind of degree n

$$T_n(\Omega) = \begin{cases} \cos(n \cos^{-1} \Omega) \quad , & |\Omega| \leqslant 1 \\ \cosh(n \cosh^{-1} \Omega) \quad , & |\Omega| \geqslant 1 \end{cases} \tag{2-4}$$

TABLE I

POLES OF BUTTERWORTH TRANSFER FUNCTION [Zeros of $Q_n(\lambda)$]

| n = 2 | n = 3 | n = 4 | n = 5 |
|---|---|---|---|
| −0.70711 ± i 0.70711 | −1.00000 | −0.92388 ± i 0.38268 | −1.00000 |
| | −0.50000 ± i 0.86603 | −0.38268 ± i 0.92388 | −0.80902 ± i 0.58779 |
| | | | −0.30902 ± i 0.95106 |

TABLE II

COEFFICIENTS OF $Q_n(\lambda)$ FOR A BUTTERWORTH TRANSFER FUNCTION

$$Q_n(\lambda) = \lambda^n + q_{n-1}\lambda^{n-1} + \ldots + q_1\lambda + 1$$

| n | $q_1$ | $q_2$ | $q_3$ | $q_4$ |
|---|---|---|---|---|
| 2 | 1.414214 | | | |
| 3 | 2.000000 | 2.000000 | | |
| 4 | 2.613126 | 3.414214 | 2.613126 | |
| 5 | 3.236068 | 5.236068 | 5.236068 | 3.236068 |

and the peak-to-peak ripple $\gamma$ in decibels is

$$\gamma = 10 \log(1 + \epsilon^2) \quad . \tag{2-5}$$

A particular Chebyshev transfer function is specified by two parameters n and $\gamma$ (or $\epsilon$), and will be designated by the notation CH $n - \gamma$. Thus, CH 5 – 0.03 refers to a five-pole Chebyshev transfer function whose ripple is 0.03 dB. A Chebyshev transfer function is

$$t_{CH}(\lambda) = \frac{1}{\epsilon 2^{n-1} Q_n(\lambda)} \quad . \tag{2-6}$$

The zeros of $Q_n(\lambda)$ and its coefficients are given in Tables III through V.

Bessel filters are characterized by having a group delay that approximates 1 second in a maximally flat sense. The transfer function is

$$t_{BE}(\lambda) = \frac{(2n)!}{2^n n!} \frac{1}{y_n(\lambda)} = \frac{q_0}{Q_n(\lambda)} \tag{2-7}$$

where $y_n(\lambda) = \lambda^n w_n(1/\lambda)$ and $w_n(x)$ is a Bessel polynomial. The zeros and coefficients of this $Q_n(\lambda)$ are given in Tables VI and VII. The transfer function is thus completely specified by the parameter n; both the bandwidth and the band over which the time delay is substantially constant increases with n. We will designate a particular characteristic by the notation BE n. Thus, BE 3 refers to a three-pole Bessel transfer function.

For a full discussion of the characteristics of these three transfer functions see Chap. 11 of Weinberg's book[7] and the references cited there.

## B. BANDWIDTH

Assume that a bandpass filter is required to have a bandwidth $W_\alpha = 2\pi B_\alpha$ rad/sec between the points at which the transmission is a $\alpha$ dB below its peak value. If the reference lowpass transfer function is $\alpha$ dB below its peak value at a frequency of $\beta_\alpha$ rad/sec, then

$$w = 2\pi b = W_\alpha/\beta_\alpha = 2\pi B_\alpha/\beta_\alpha \quad . \tag{2-8}$$

TABLE III

POLES OF CHEBYSHEV TRANSFER FUNCTION [Zeros of $Q_n(\lambda)$]

| $\gamma$ (dB) | n = 2 | n = 3 | n = 4 | n = 5 | n = 6 |
|---|---|---|---|---|---|
| 0.001 | | −1.22315 ± i 2.28873<br>−2.44630 | −0.59185 ± i 1.70152<br>−1.42885 ± i 0.70479 | −0.35194 ± i 1.44144<br>−0.92139 ± i 0.89086<br>−1.13890 | −0.23457 ± i 1.30361<br>−0.64086 ± i 0.95431<br>−0.87543 ± i 0.34930 |
| 0.01 | −2.22777 ± i 2.33729 | −0.79469 ± i 1.62622<br>−1.58937 | −0.41087 ± i 1.35553<br>−0.99192 ± i 0.56148 | −0.25251 ± i 1.22820<br>−0.66109 ± i 0.75907<br>−0.81715 | −0.17147 ± i 1.15867<br>−0.46845 ± i 0.84820<br>−0.63992 ± i 0.31046 |
| 0.03 | −1.66227 ± i 1.80642 | −0.63517 ± i 1.40011<br>−1.27034 | −0.33740 ± i 1.23169<br>−0.81455 ± i 0.51018 | −0.21011 ± i 1.15007<br>−0.55007 ± i 0.71078<br>−0.67992 | −0.14372 ± i 1.10486<br>−0.39265 ± i 0.80881<br>−0.53638 ± i 0.29605 |
| 0.10 | −1.18618 ± i 1.38095 | −0.48470 ± i 1.20616<br>−0.96941 | −0.26416 ± i 1.12261<br>−0.63773 ± i 0.46500 | −0.16653 ± i 1.08037<br>−0.43599 ± i 0.66771<br>−0.53891 | −0.11469 ± i 1.05652<br>−0.31335 ± i 0.77343<br>−0.42804 ± i 0.28309 |
| 0.30 | −0.84716 ± i 1.10348 | −0.36464 ± i 1.07186<br>−0.72928 | −0.20260 ± i 1.04536<br>−0.48912 ± i 0.43300 | −0.12890 ± i 1.03048<br>−0.33746 ± i 0.63687<br>−0.41713 | −0.08922 ± i 1.02171<br>−0.24376 ± i 0.74794<br>−0.33298 ± i 0.27377 |
| 1.0 | −0.54887 ± i 0.89513 | −0.24709 ± i 0.96600<br>−0.49417 | −0.13954 ± i 0.98338<br>−0.33687 ± i 0.40733 | −0.08946 ± i 0.99011<br>−0.23421 ± i 0.61192<br>−0.28949 | −0.06218 ± i 0.99341<br>−0.16988 ± i 0.72723<br>−0.23206 ± i 0.26618 |

TABLE IV

NUMERICAL COEFFICIENTS OF $Q_n(\lambda)$ FOR A CHEBYSHEV TRANSFER FUNCTION

$$Q_n(\lambda) = \lambda^n + q_{n-1}\lambda^{n-1} + \ldots + q_1\lambda + q_0$$

| $\gamma$ (dB) | n | $q_0$ | $q_1$ | $q_2$ | $q_3$ | $q_4$ |
|---|---|---|---|---|---|---|
| 0.001 | 3 | 16.474307 | 12.718804 | 4.892607 | | |
| | 4 | 8.238102 | 12.279242 | 9.166493 | 4.041409 | |
| | 5 | 4.118577 | 9.553635 | 11.068605 | 8.041646 | 3.685552 |
| 0.01 | 2 | 10.425865 | 4.455534 | | | |
| | 3 | 5.206934 | 5.802197 | 3.178741 | | |
| | 4 | 2.606466 | 5.047686 | 4.935622 | 2.805574 | |
| | 5 | 1.301734 | 3.709816 | 5.248800 | 4.746274 | 2.644343 |
| 0.03 | 2 | 6.026308 | 3.324547 | | | |
| | 3 | 3.002764 | 3.977509 | 2.540673 | | |
| | 4 | 1.506577 | 3.280243 | 3.653977 | 2.303900 | |
| | 5 | 0.750691 | 2.357258 | 3.635990 | 3.670604 | 2.200275 |
| 0.10 | 2 | 3.314036 | 2.372356 | | | |
| | 3 | 1.638050 | 2.629494 | 1.938811 | | |
| | 4 | 0.828509 | 2.025500 | 2.626798 | 1.803772 | |
| | 5 | 0.409513 | 1.435558 | 2.396959 | 2.770704 | 1.743963 |
| 0.30 | 2 | 1.935344 | 1.694311 | | | |
| | 3 | 0.934821 | 1.813691 | 1.458555 | | |
| | 4 | 0.483836 | 1.282057 | 1.956934 | 1.383426 | |
| | 5 | 0.233705 | 0.919769 | 1.600987 | 2.161058 | 1.349858 |
| 1.0 | 2 | 1.102510 | 1.097734 | | | |
| | 3 | 0.491307 | 1.238409 | 0.988341 | | |
| | 4 | 0.275628 | 0.742619 | 1.453925 | 0.952811 | |
| | 5 | 0.1228267 | 0.580534 | 0.974396 | 1.688816 | 0.936820 |

TABLE V

EXPLICIT COEFFICIENTS OF $Q_n(\lambda)$ FOR A CHEBYSHEV TRANSFER FUNCTION

$$Q_n(\lambda) = \lambda^n + q_{n-1}\lambda^{n-1} + \ldots + q_1\lambda + q_0$$

| n | Coefficients |
|---|---|
| 2 | $q_0 = 1/2 + \sinh^2\phi \quad , \quad q_1 = \sqrt{2}\sinh\phi$ |
| 3 | $q_0 = (3/4 + \sinh^2\phi)\sinh\phi \quad , \quad q_1 = 3/4 + 2\sinh\phi \quad , \quad q_2 = 2\sinh\phi$ |
| 4 | $q_0 = 1/8 + \sinh^2\phi + \sinh^4\phi \quad , \quad q_1 = \sqrt{2(2+\sqrt{2})}\left(1 - \frac{1}{2\sqrt{2}} + \sinh^2\phi\right)\sinh\phi$ |
| | $q_2 = 1 + (2+\sqrt{2})\sinh^2\phi \quad , \quad q_3 = \sqrt{2(2+\sqrt{2})}\sinh\phi$ |
| 5 | $q_0 = (\sinh^4\phi + 5/4\sinh^2\phi + 5/16)\sinh\phi$ |
| | $q_1 = (1+\sqrt{5})\sinh^4\phi + (\frac{5+3\sqrt{5}}{4})\sinh^2\phi + 5/16$ |
| | $q_2 = (3+\sqrt{5})\sinh^3\phi + (\frac{5+3\sqrt{5}}{4})\sinh\phi$ |
| | $q_3 = (3+\sqrt{5})\sinh^2\phi + 5/4$ |
| | $q_4 = (1+\sqrt{5})\sinh\phi$ |
| | where $\phi = 1/n\,\sinh^{-1} 1/\epsilon = 1/n\,\coth^{-1}(10^{\gamma/20})$ |
| | and $\gamma = 10\log(1+\epsilon^2)$ . |
| | A general and simpler expression for $q_0$ is $q_0 = \begin{cases} 1/\epsilon 2^{n-1} & , \text{ n odd} \\ \sqrt{1+\epsilon^2}/\epsilon 2^{n-1} & , \text{ n even} \end{cases}$ . |

| TABLE VI<br>POLES OF BESSEL TRANSFER FUNCTION [Zeros of $Q_n(\lambda)$] | | | |
|---|---|---|---|
| n = 2 | n = 3 | n = 4 | n = 5 |
| $-1.50000 \pm i\,0.86603$ | $-2.32219$ | $-2.89621 \pm i\,0.86723$ | $-3.64674$ |
| | $-1.83891 \pm i\,1.75438$ | $-2.10379 \pm i\,2.65742$ | $-3.35196 \pm i\,1.74266$ |
| | | | $-2.32467 \pm i\,3.57102$ |

TABLE VII

COEFFICIENTS OF $Q_n(\lambda)$ FOR A BESSEL TRANSFER FUNCTION

$$Q_n(\lambda) = \lambda^n + q_{n-1}\,\lambda^{n-1} + \ldots + q_1\lambda + q_0$$

| n | $q_0$ | $q_1$ | $q_2$ | $q_3$ | $q_4$ |
|---|---|---|---|---|---|
| 2 | 3 | 3 | | | |
| 3 | 15 | 15 | 6 | | |
| 4 | 105 | 105 | 45 | 10 | |
| 5 | 945 | 945 | 420 | 105 | 15 |

TABLE VIII

HALF-POWER BANDWIDTH OF CHEBYSHEV TRANSFER FUNCTIONS, $\beta_3$

| Ripple, $\gamma$ (dB) | $\beta_3 = \cosh\left(\frac{1}{n}\cosh^{-1}\frac{1}{\epsilon}\right)$ rad/sec | | | | |
|---|---|---|---|---|---|
| | n = 2 | n = 3 | n = 4 | n = 5 | n = 6 |
| 0.001 | 5.783493 | 2.642704 | 1.841670 | 1.515589 | 1.349575 |
| 0.01 | 3.303619 | 1.877180 | 1.466905 | 1.291218 | 1.199412 |
| 0.03 | 2.550594 | 1.615244 | 1.332403 | 1.208783 | 1.143513 |
| 0.10 | 1.943219 | 1.388995 | 1.213099 | 1.134718 | 1.092931 |
| 0.30 | 1.539364 | 1.229063 | 1.126802 | 1.080553 | 1.055714 |
| 1.0 | 1.217626 | 1.094868 | 1.053002 | 1.033815 | 1.023442 |

For a lowpass filter, w is similarly defined. Assume that a lowpass filter passes all frequencies out to $W_\alpha = 2\pi B_\alpha$ rad/sec and there has a transmission that is $\alpha$ dB below the peak value. If the reference lowpass transfer function is $\alpha$ dB below its peak value at a frequency of $\beta_\alpha$ rad/sec, then Eq. (2-8) is the proper definition for w in the lowpass case also.

Thus, if a Butterworth filter is required to have a 3-dB bandwidth of $W_3$ rad/sec, then $\beta_\alpha = \beta_3 = 1$ and $w = W_3$ with the lowpass normalization used here. If a Chebyshev filter is required to have an equal-ripple bandwidth of $W_\gamma$ rad/sec, then $\beta_\alpha = \beta_\gamma = 1$ and $w = W_\gamma$. For Chebyshev filters another often used bandwidth specification is the 3-dB bandwidth; for the calculation of w in this case, $\beta_3$ for the Chebyshev characteristic is required. It is given by

$$\beta_3 = \cosh\left(\frac{1}{n} \cosh^{-1} \frac{1}{\epsilon}\right) \quad . \tag{2-9}$$

In Table VIII, values of $\beta_3$ are given for n = 2 through 6 and for $\gamma = 10 \log(1 + \epsilon^2) = 0.001$, 0.01, 0.03, 0.10, 0.30 and 1.0 dB.

A Bessel filter (as normalized here) does not have a uniform bandwidth; its bandwidth for $\alpha$ up to 7.5 dB can be found on the expanded attenuation plot in Fig. 9.

For bandwidth specifications other than those discussed above, $\beta_\alpha$ can be found with little difficulty for Butterworth and Chebyshev filters by direct calculation using Eqs. (2-1) and (2-3). For Bessel filters one must resort to tables of spherical Bessel functions as discussed by Weinberg[7].

## C. FREQUENCY CHARACTERISTICS

The attenuation $A_L(\Omega)$, phase $\varphi_L(\Omega)$, and group delay $\tau_L(\Omega)$ of Butterworth, Chebyshev, and Bessel transfer functions [Eqs. (2-2), (2-6), and (2-7), respectively] are given in Figs. 1 through 23.

## D. TRANSIENT RESPONSES

Figures 24 through 39 show the unit impulse response, $h_L(t) = \mathcal{L}^{-1}[t(\lambda)]$, and the unit step response, $g_L(t) = \mathcal{L}^{-1}[\lambda^{-1} t(\lambda)]$, of the various transfer functions where $\mathcal{L}^{-1}$ indicates the inverse Laplace transformation.

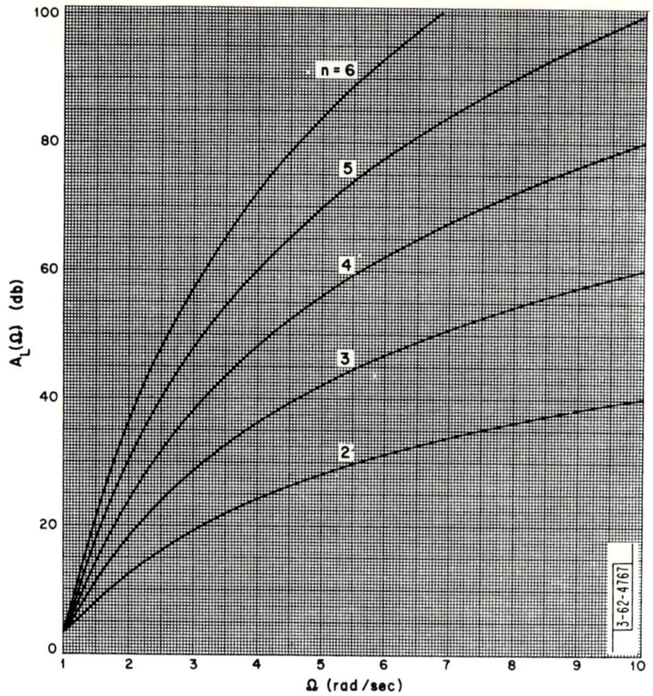


Fig. 1. Butterworth attenuation characteristics.

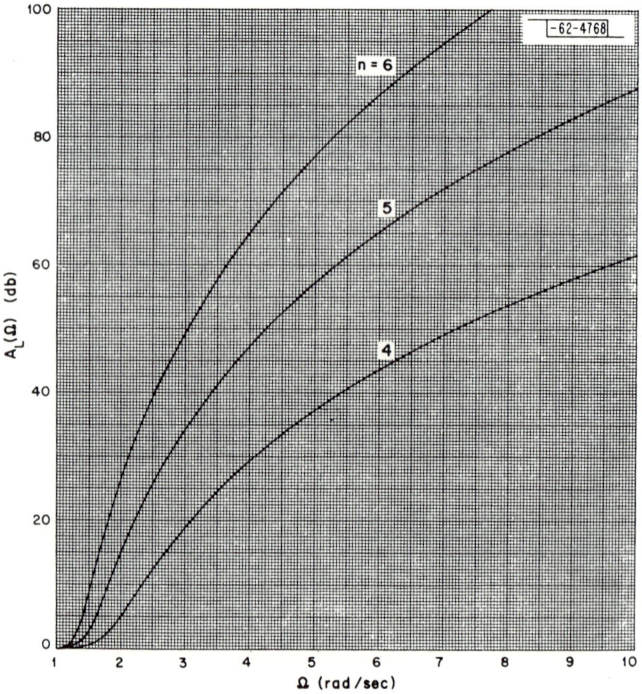


Fig. 2. Chebyshev attenuation characteristics for 0.001-dB ripple.

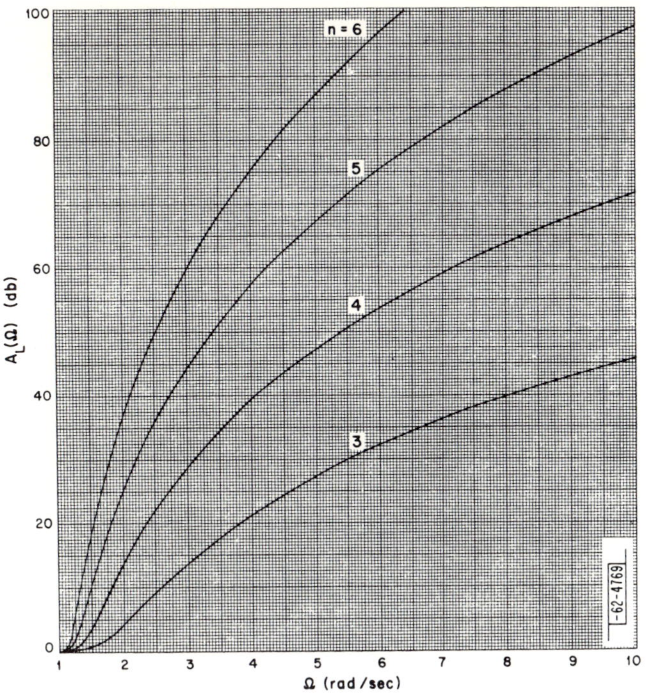


Fig. 3. Chebyshev attenuation characteristics for 0.01-dB ripple.

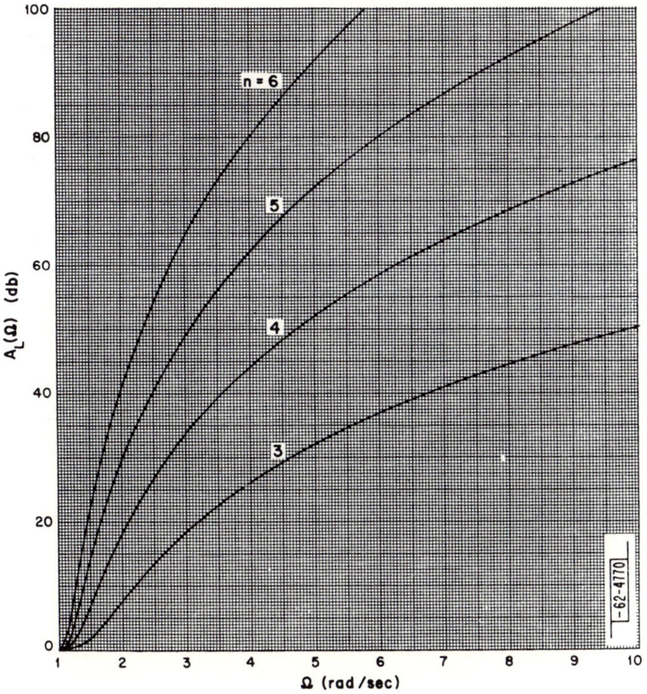


Fig. 4. Chebyshev attenuation characteristics for 0.03-dB ripple.

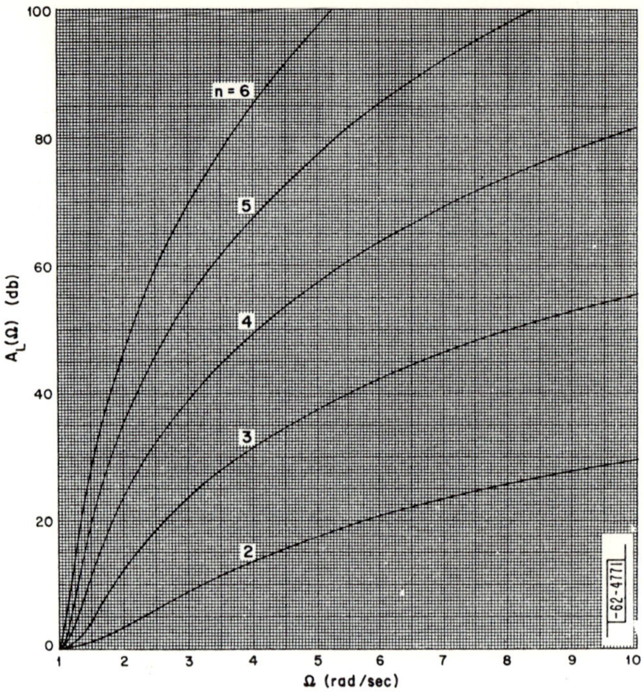


Fig. 5. Chebyshev attenuation characteristics for 0.1-dB ripple.

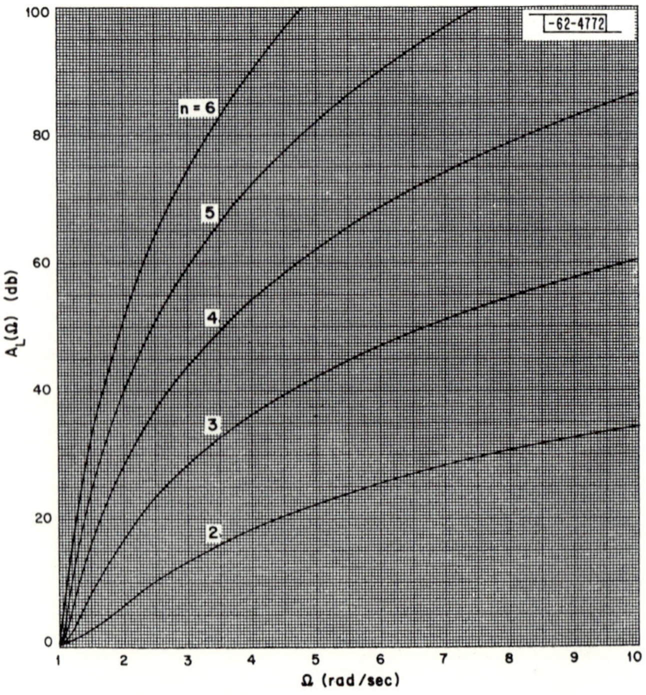


Fig. 6. Chebyshev attenuation characteristics for 0.3-dB ripple.

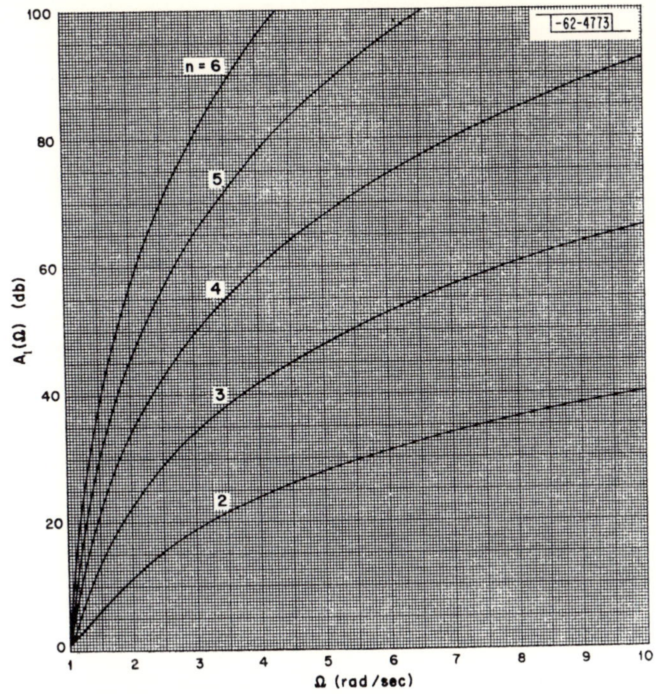


Fig. 7. Chebyshev attenuation characteristics for 1.0-dB ripple.

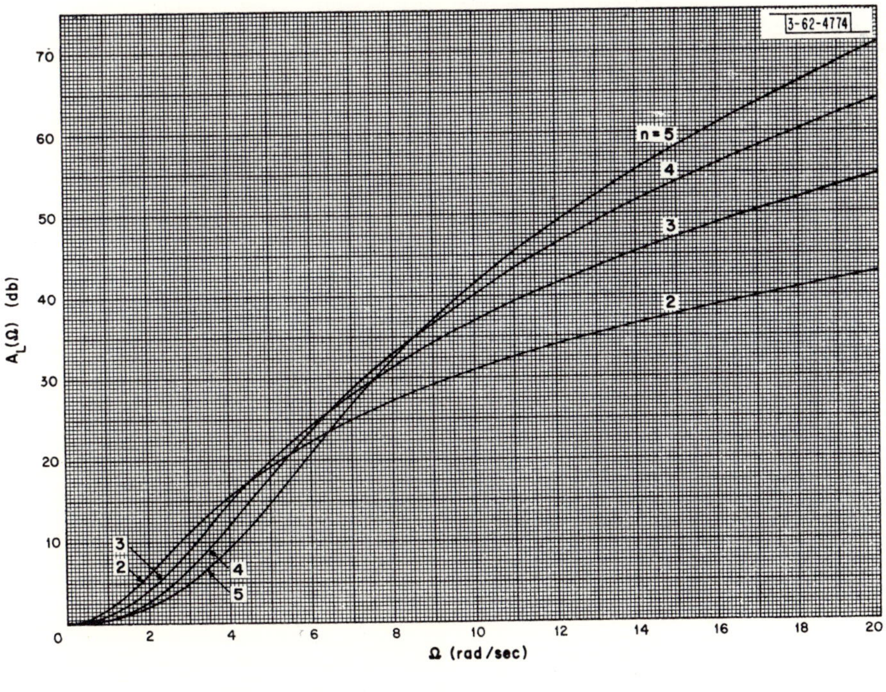


Fig. 8. Bessel attenuation characteristics.

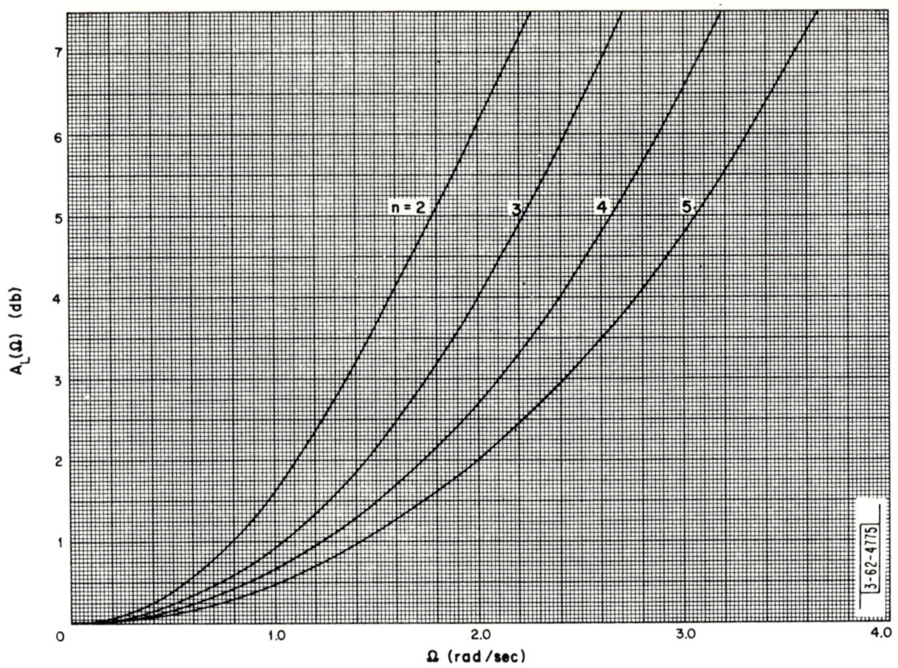


Fig. 9. Bessel attenuation characteristics for the pass band.

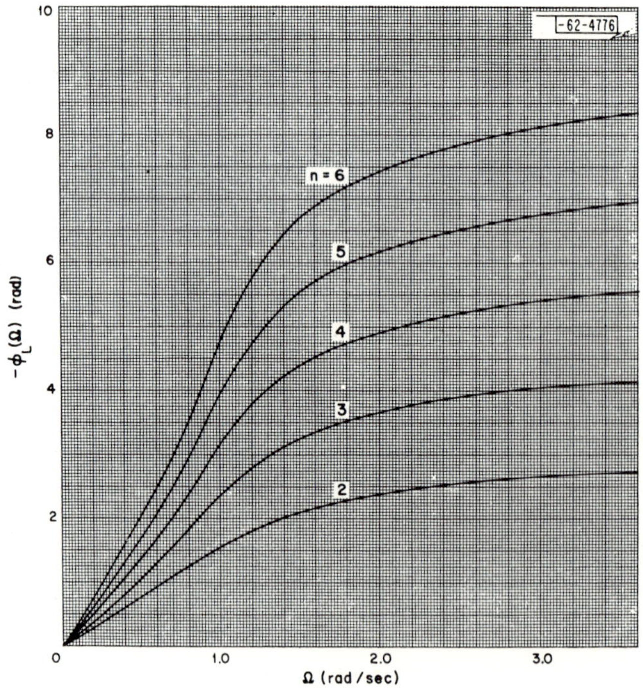


Fig. 10. Butterworth phase characteristics.

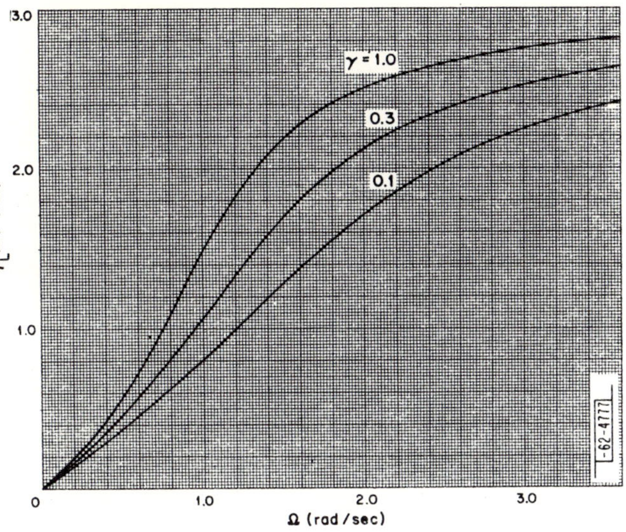


Fig. 11. Two-pole Chebyshev phase characteristics.

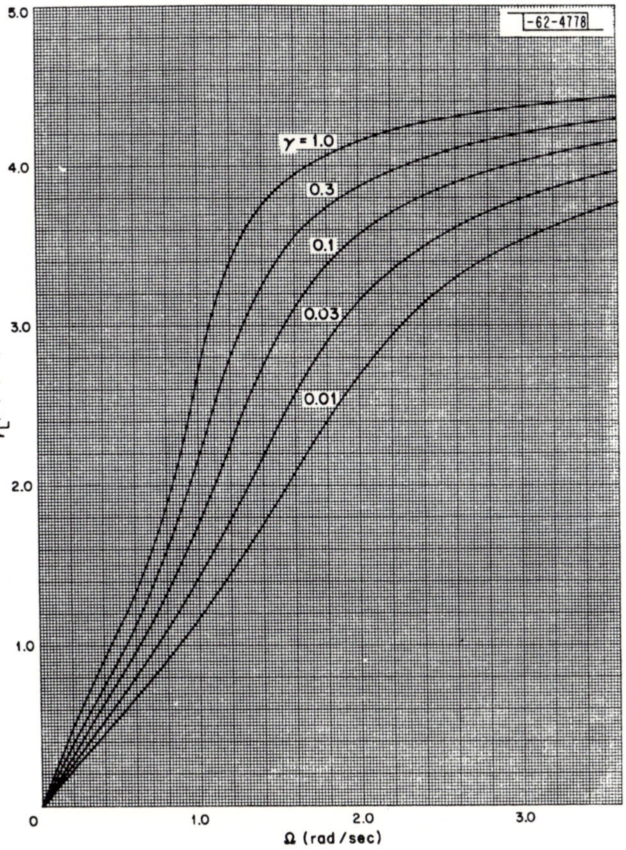


Fig. 12. Three-pole Chebyshev phase characteristics.

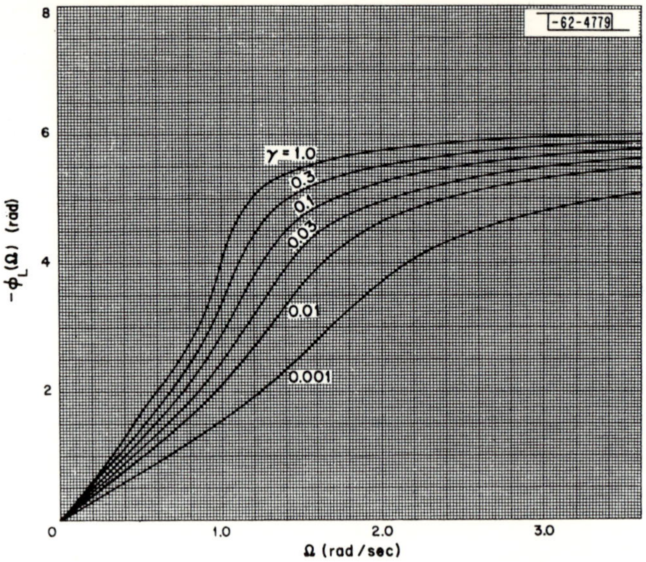


Fig. 13. Four-pole Chebyshev phase characteristics.

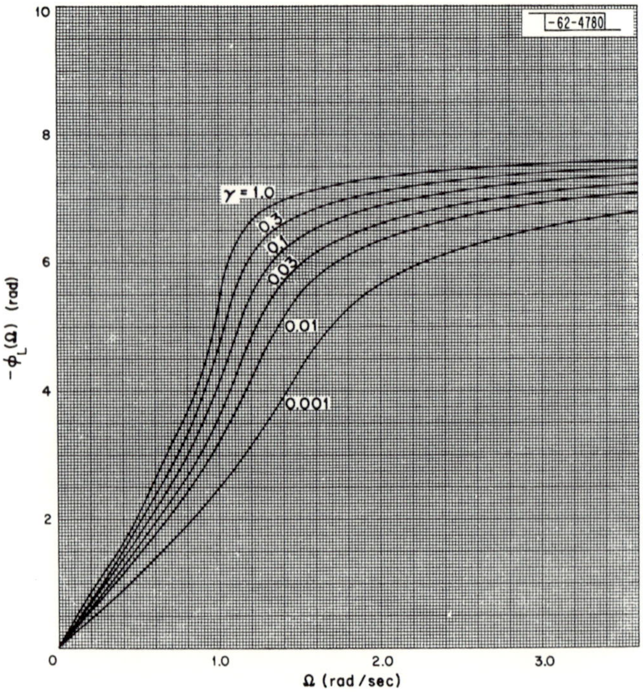


Fig. 14. Five-pole Chebyshev phase characteristics.

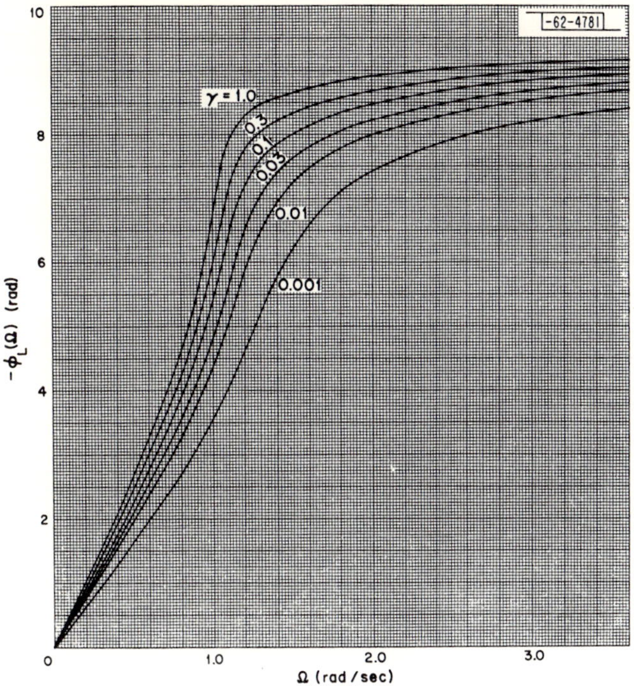


Fig. 15. Six-pole Chebyshev phase characteristics.

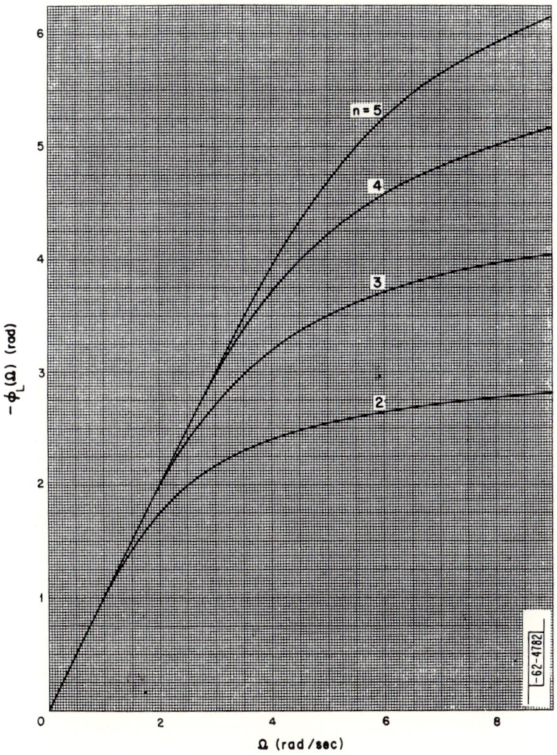


Fig. 16. Bessel phase characteristics.

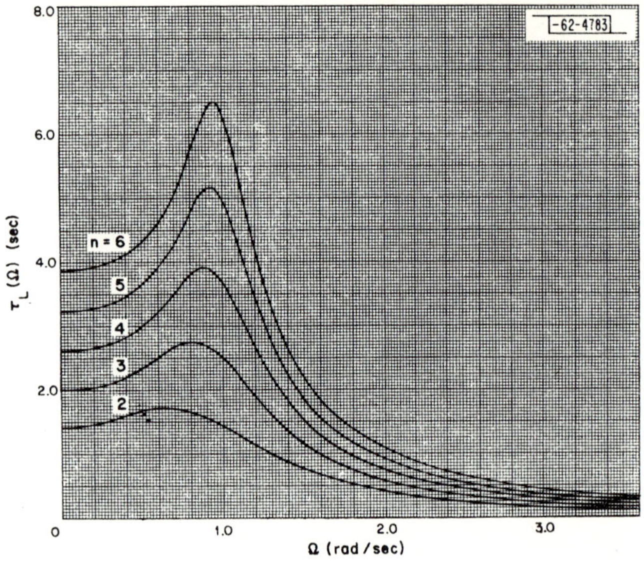


Fig. 17. Butterworth group delay characteristics.

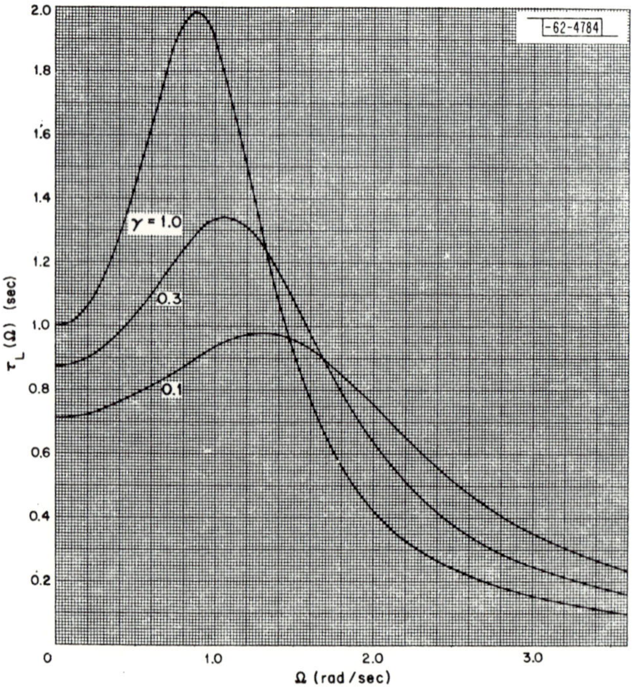


Fig. 18. Two-pole Chebyshev group delay characteristics.

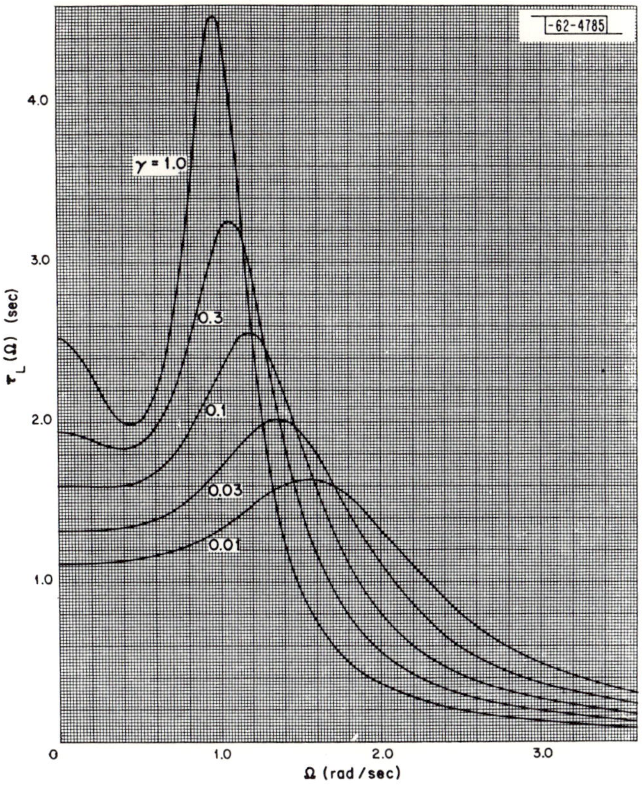


Fig. 19. Three-pole Chebyshev group delay characteristics.

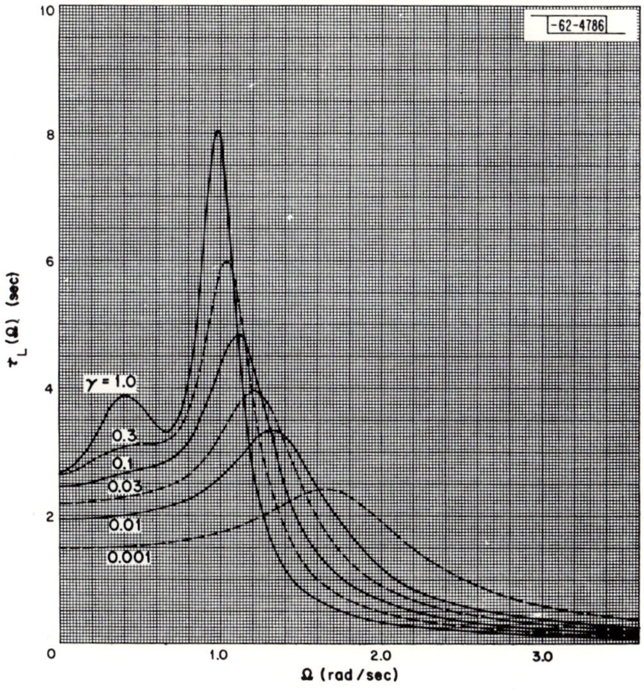


Fig. 20. Four-pole Chebyshev group delay characteristics.

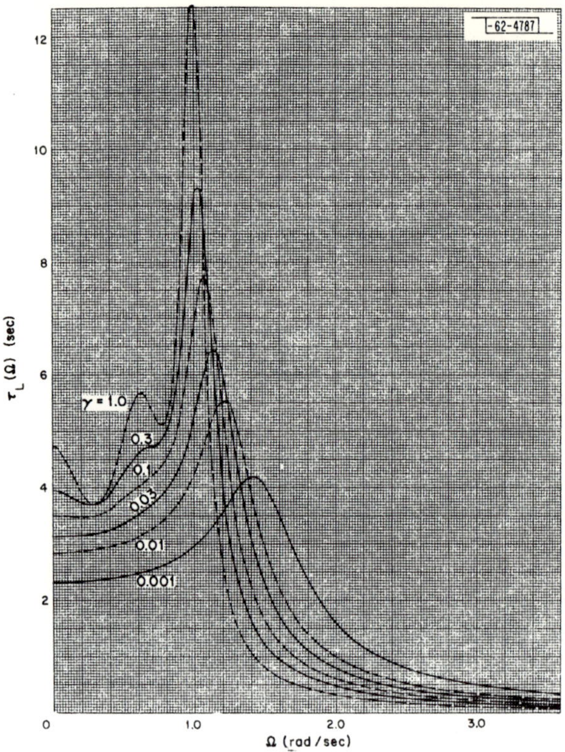


Fig. 21. Five-pole Chebyshev group delay characteristics.

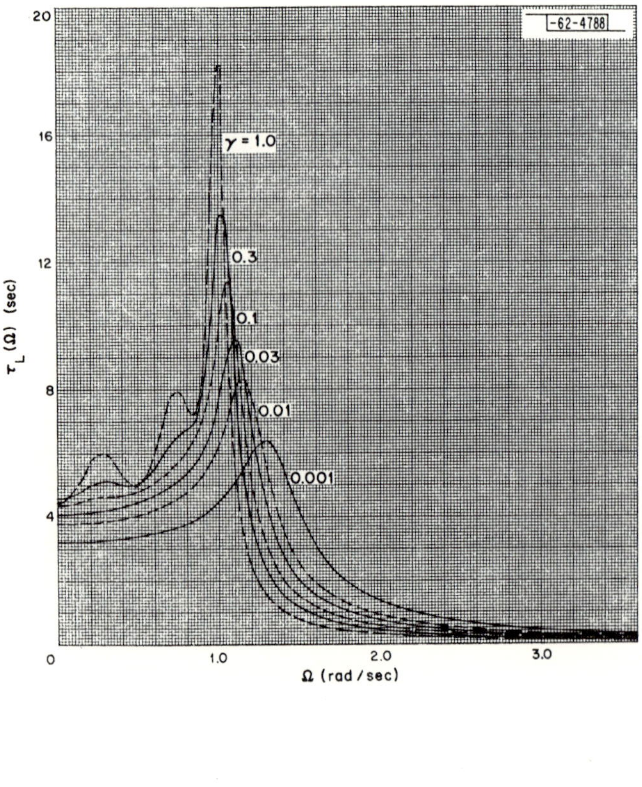


Fig. 22. Six-pole Chebyshev group delay characteristics.

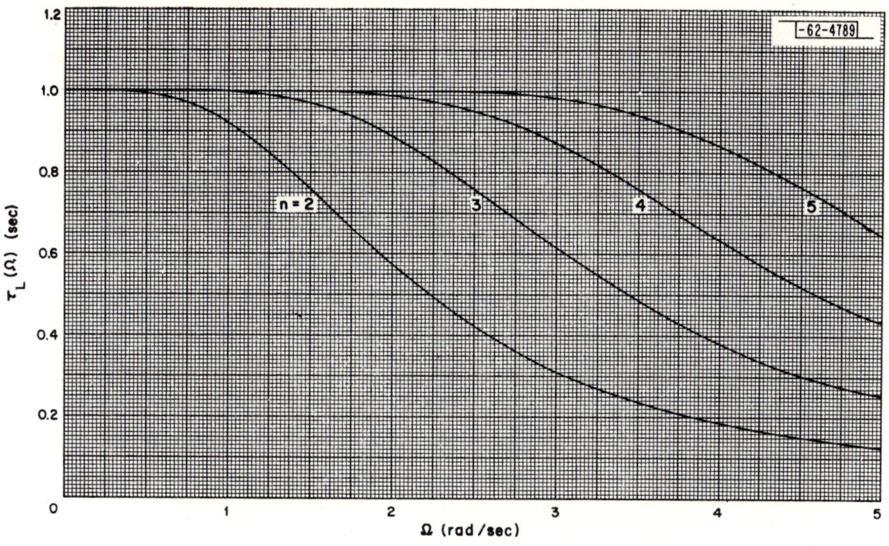


Fig. 23. Bessel group delay characteristics.

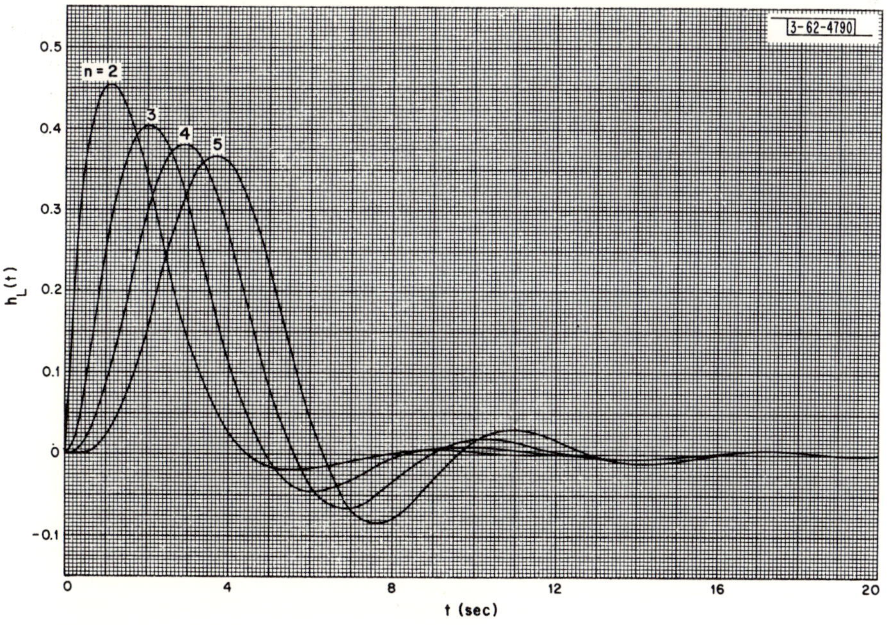


Fig. 24. Impulse response of Butterworth filters.

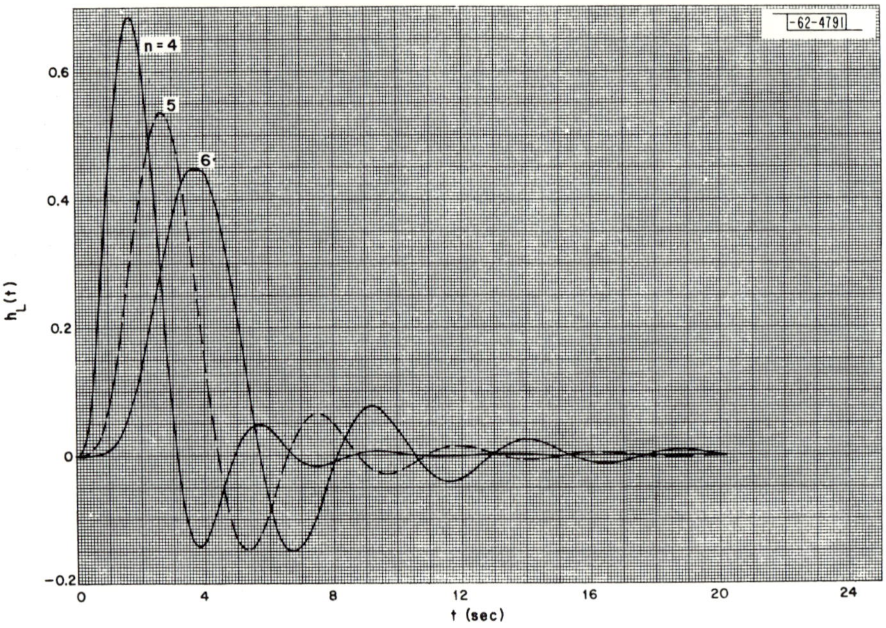


Fig. 25. Impulse response of Chebyshev filters with 0.001-dB ripple.

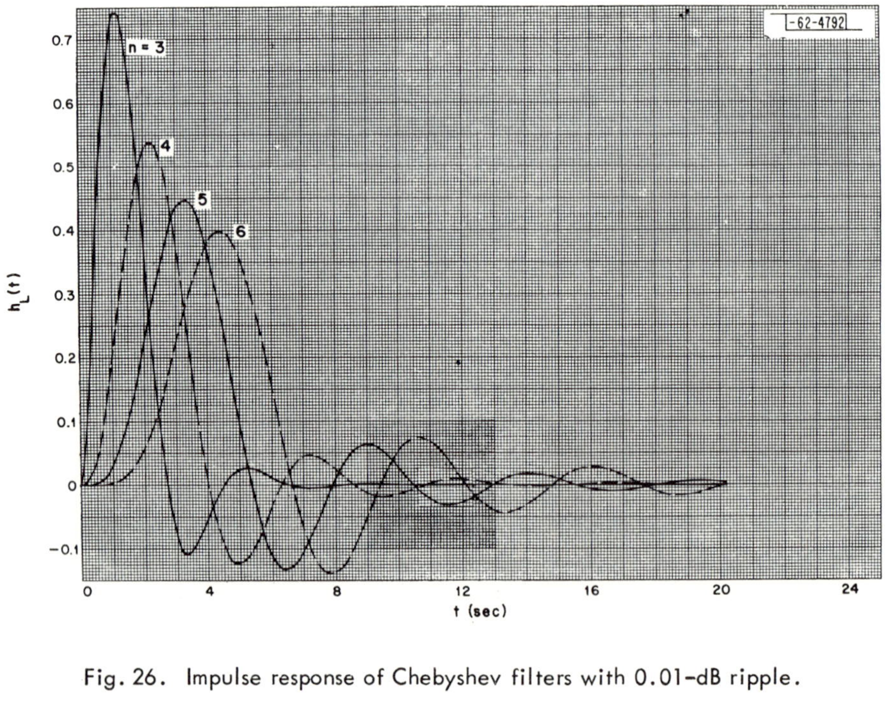


Fig. 26. Impulse response of Chebyshev filters with 0.01-dB ripple.

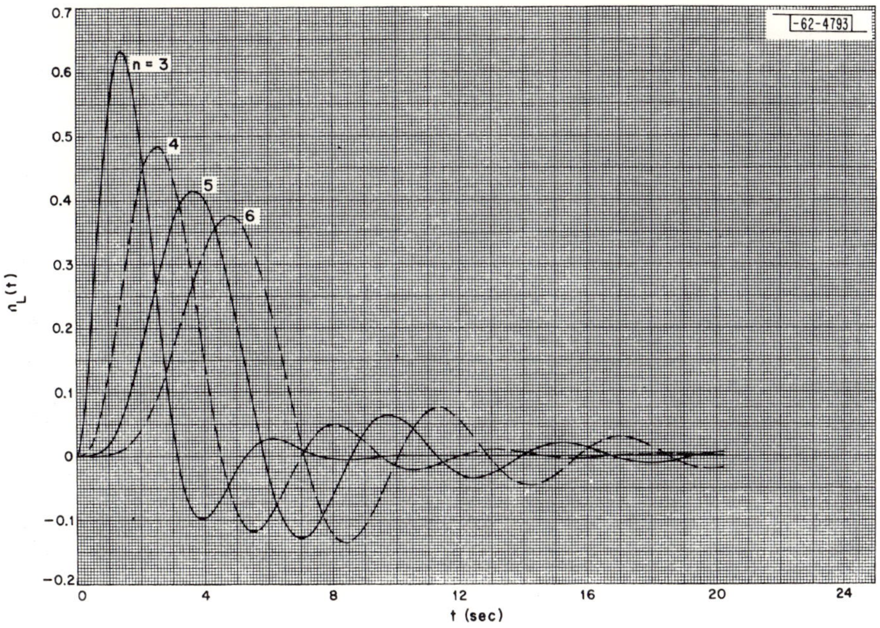


Fig. 27. Impulse response of Chebyshev filters with 0.03-dB ripple.

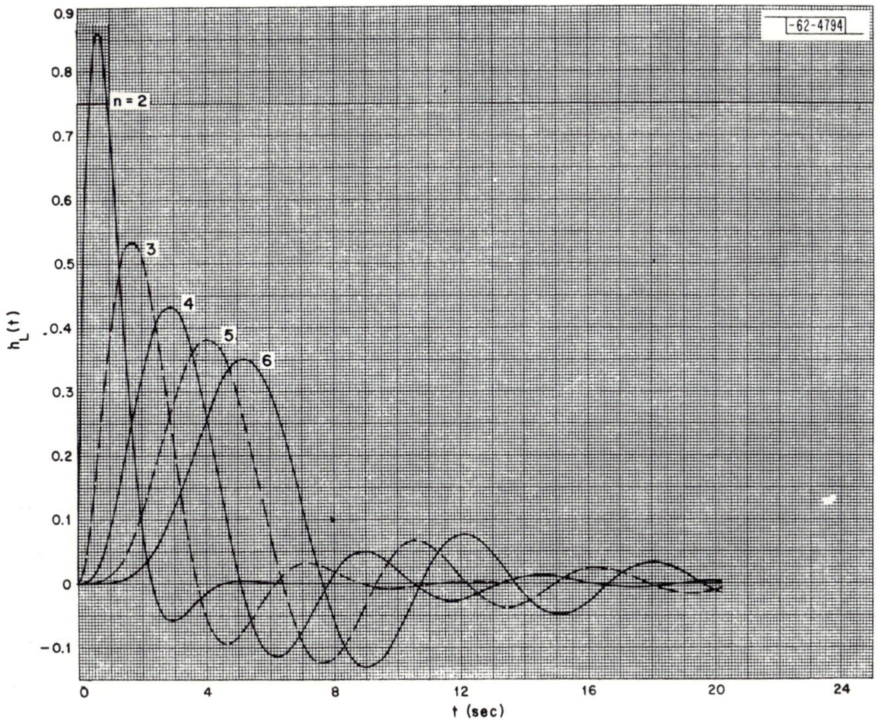


Fig. 28. Impulse response of Chebyshev filters with 0.1-dB ripple.

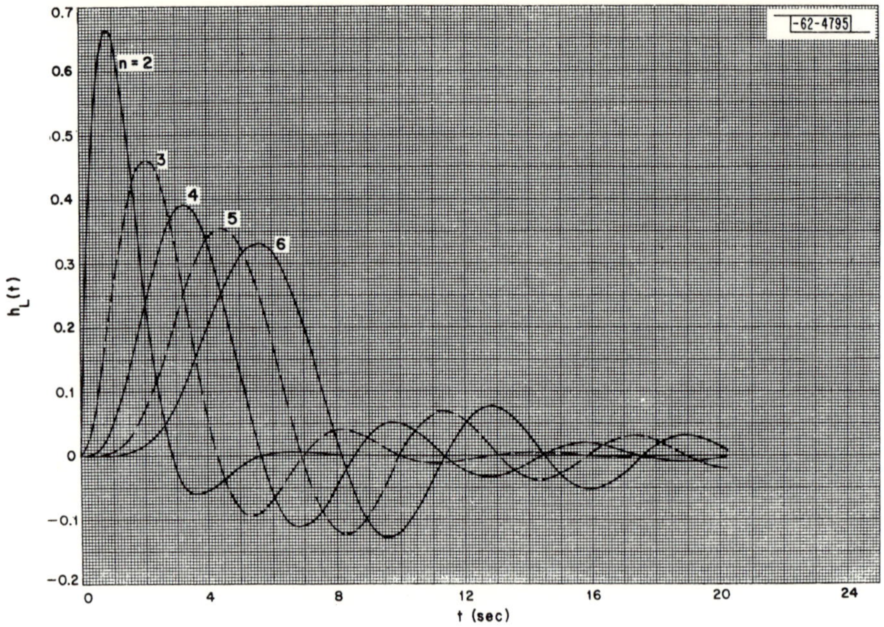


Fig. 29. Impulse response of Chebyshev filters with 0.3-dB ripple.

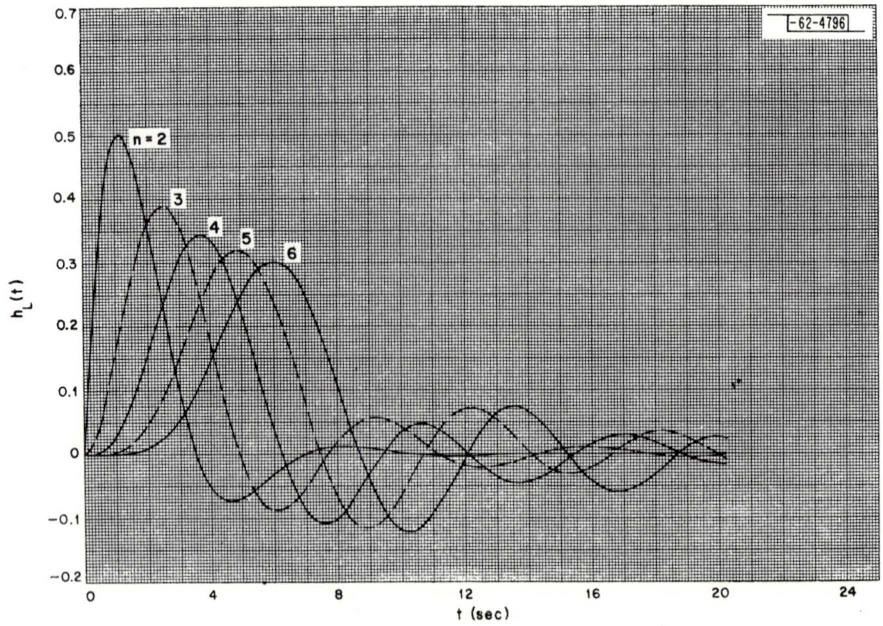


Fig. 30. Impulse response of Chebyshev filters with 1.0-dB ripple.

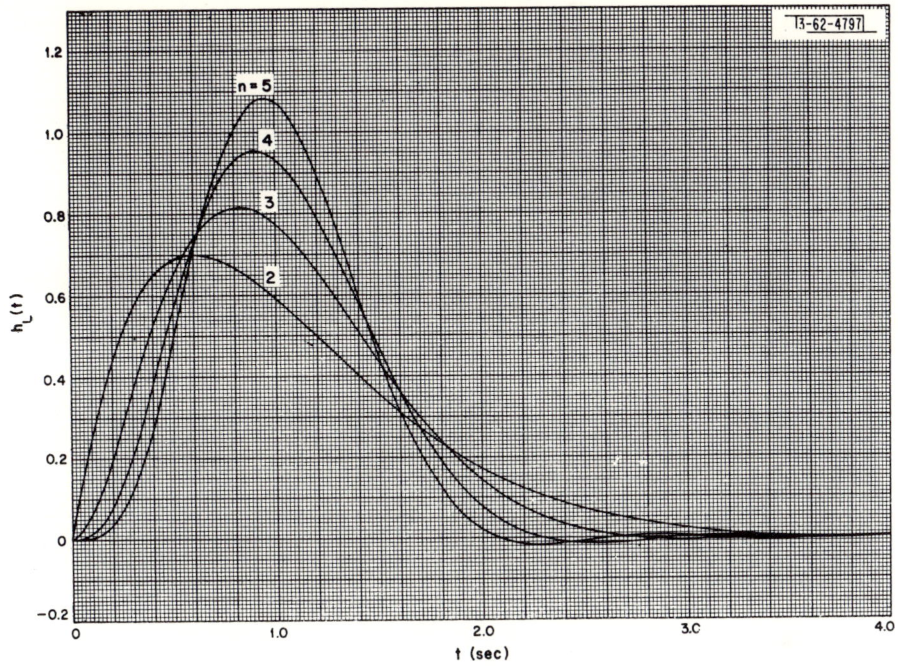


Fig. 31. Impulse response of Bessel filters.

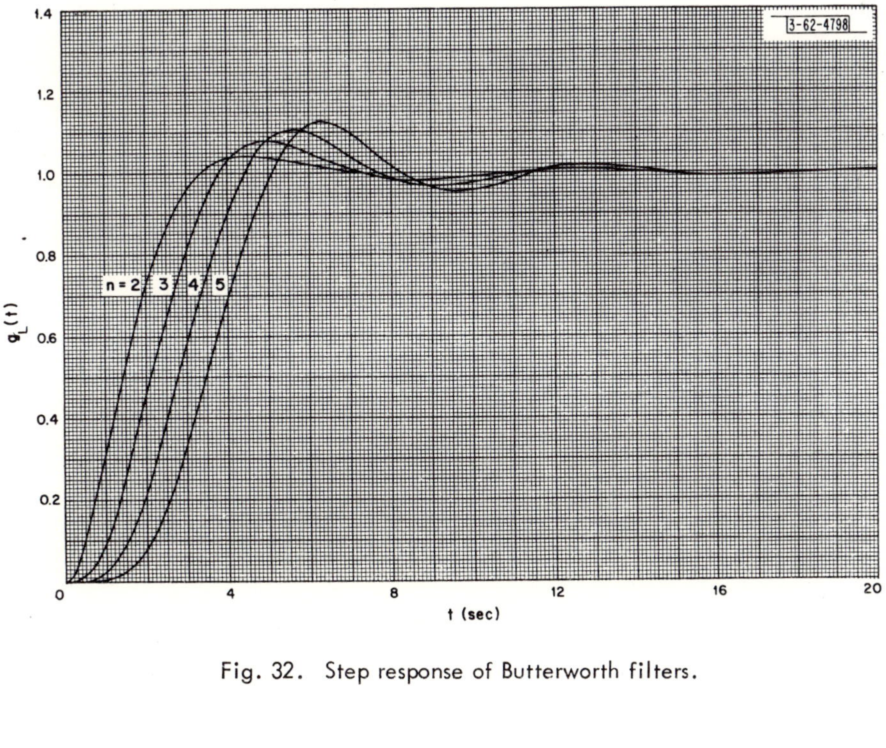


Fig. 32. Step response of Butterworth filters.

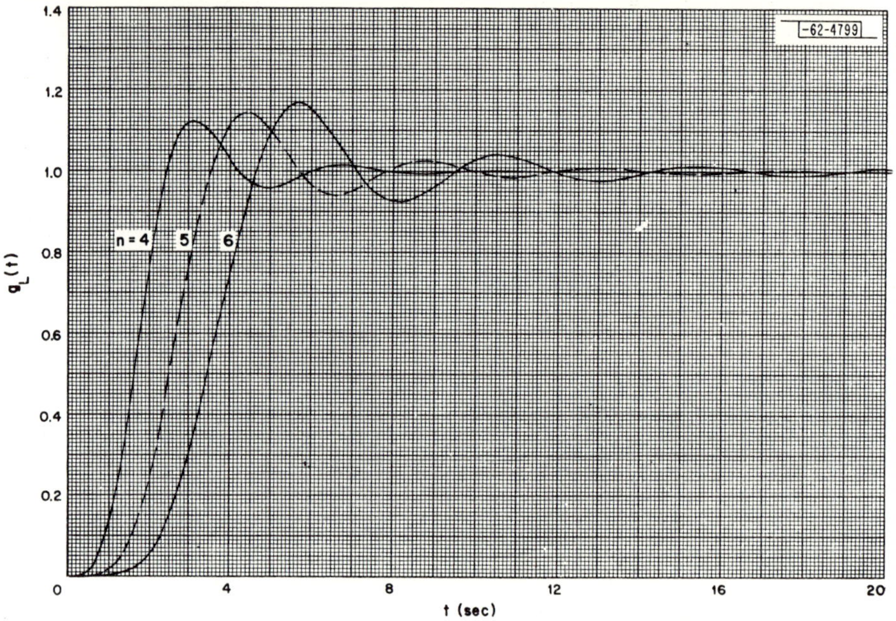


Fig. 33. Step response of Chebyshev filters with 0.001-dB ripple.

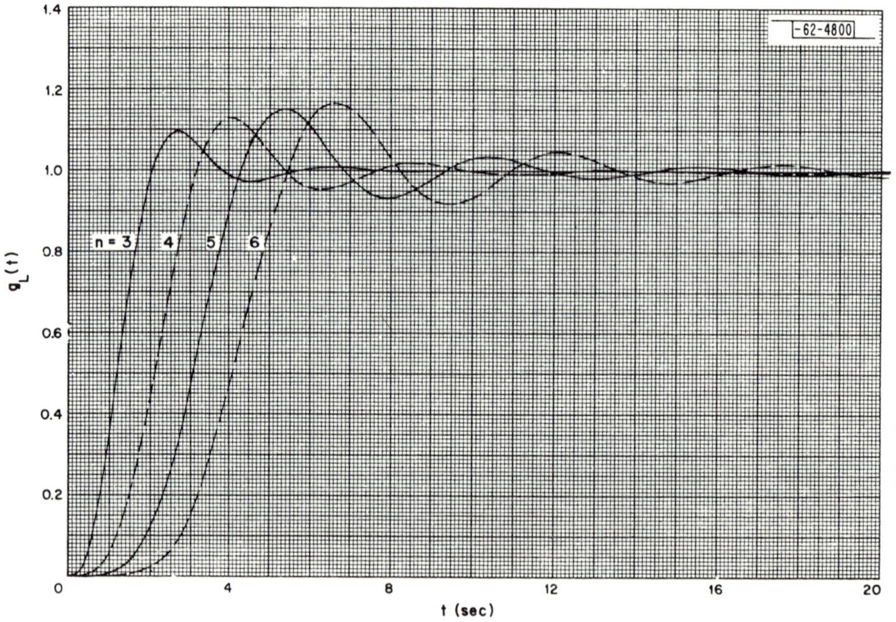


Fig. 34. Step response of Chebyshev filters with 0.01-dB ripple.

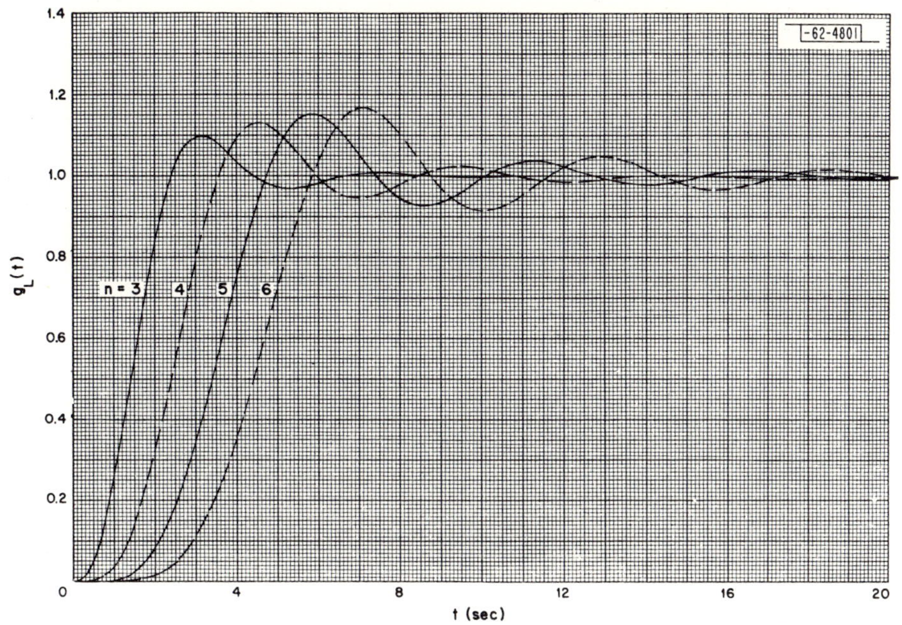


Fig. 35. Step response of Chebyshev filters with 0.03-dB ripple.

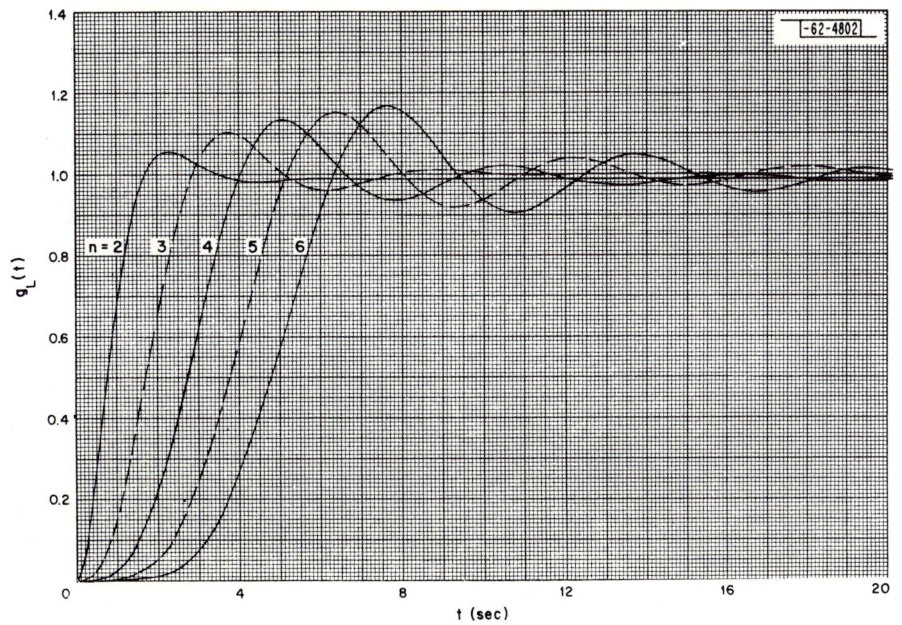


Fig. 36. Step response of Chebyshev filters with 0.1-dB ripple.

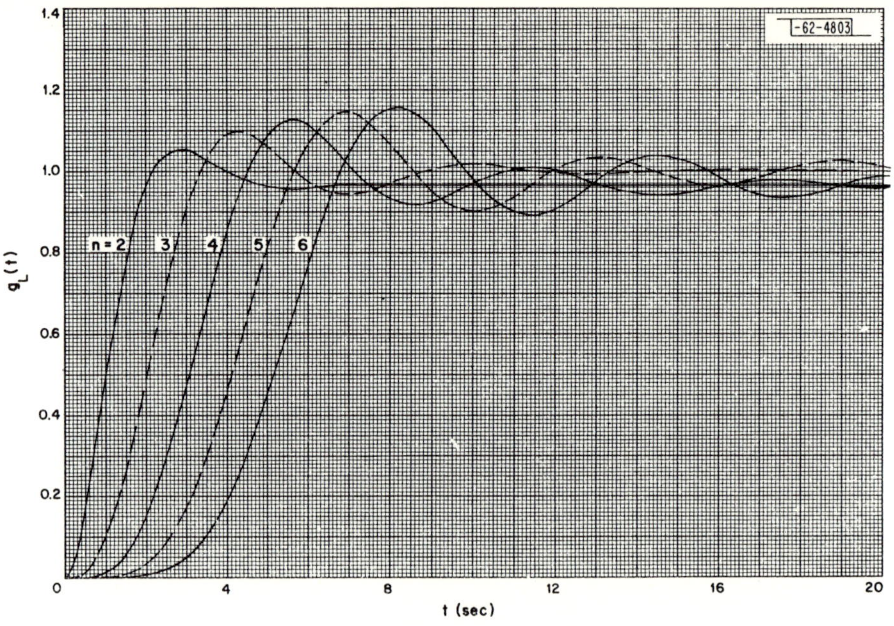


Fig. 37. Step response of Chebyshev filters with 0.3-dB ripple.

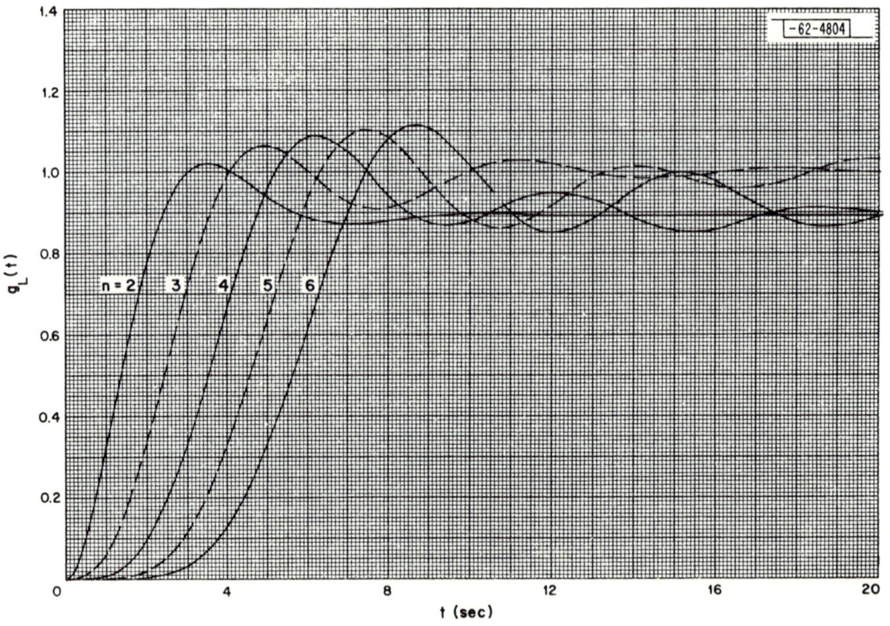


Fig. 38. Step response of Chebyshev filters with 1.0-dB ripple.

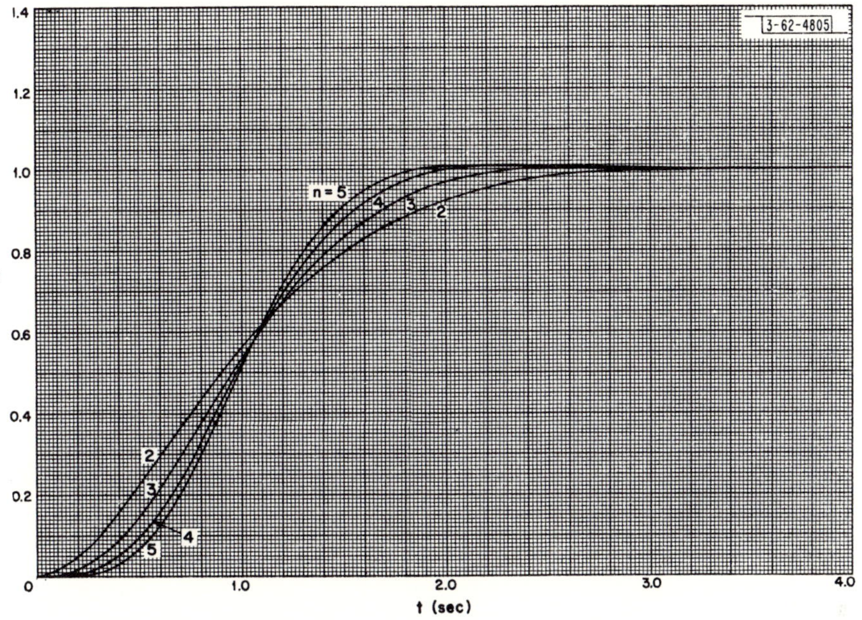


Fig. 39. Step response of Bessel filters.

## III. GUIDE TO DESIGN OF UNIFORMLY LOSSY FILTERS

In this section we outline the design of narrow bandpass, lowpass, and wide bandpass filter circuits based on the normalized design parameters. For narrow bandpass filters, three circuit configurations are considered, and for lowpass and wide bandpass filters one configuration for each is considered. In all instances, simple relationships between the circuit element values and the normalized design parameters are given.

Normalized design parameters are the coupling coefficients between adjacent parts of the network ($k_{j,\,j+1}$), the parasitic (or unloaded) dissipation factor of the uniformly lossy circuit elements a, and the loaded dissipation factors at the source and load ports of the filter (d and $\delta$, respectively). The information is presented in the following form: families of curves of the k's and $\delta$ as functions of d are given, with a as the parameter of the families. From this information and the desired bandwidth and center frequency, the actual circuit parameters can be determined. Also given in the same format are curves of the normalized gain parameter $\Gamma$, from which the transfer function of a filter at its center frequency is determined. Since $\Gamma$ gives the gain of bandpass filters at their center frequency and that of lowpass filters at zero frequency, this is also the peak value for Butterworth, Bessel, and odd-order Chebyshev filters. For even-order Chebyshev filters, the gain at a valley in the pass band is obtained, which must be multiplied by $\sqrt{1+\epsilon^2}$ to obtain the maximum gain. Values of $\sqrt{1+\epsilon^2}$ are given in Table IX.

TABLE IX

VALUES OF VARIOUS USEFUL FUNCTIONS OF $\epsilon$

| $\gamma$ | $\epsilon^2 = 10^{\gamma/10} - 1$ | $\sqrt{1+\epsilon^2} = 10^{\gamma/20}$ | $\epsilon$ |
|---|---|---|---|
| 0.001 | 0.00023 028502 | 1.00011 51358 | 0.01517 51448 |
| 0.01 | 0.00230 52381 | 1.00115 19555 | 0.04801 2895 |
| 0.03 | 0.00693 16689 | 1.00345 985 | 0.08325 6644 |
| 0.10 | 0.02329 29923 | 1.01157 945 | 0.15262 042 |
| 0.30 | 0.07151 93052 | 1.03514 217 | 0.26743 094 |
| 1.0 | 0.25892 54118 | 1.12201 845 | 0.50884 714 |

The design is based on reducing the actual filter specifications to those of an equivalent lowpass filter of unity bandwidth. This is accomplished by frequency transformations between the lowpass frequency $\Omega$ and the actual frequency $\omega$ that are discussed below. By using these transformations the attenuation, phase, and delay of the actual filter can be found from the corresponding characteristics of the equivalent lowpass filter. Likewise, transient responses of lowpass and narrow bandpass filters can be found from those of the equivalent lowpass filter.

### A. NARROW BANDPASS FILTERS

For narrow bandpass filters the normalizing factor $\rho$ is the ratio of bandwidth ($w = 2\pi b$ rad/sec) to the center frequency ($\omega_0 = 2\pi f_0$ rad/sec):

$$\rho = \frac{w}{\omega_0} = \frac{b}{f_0} \quad . \tag{3-1}$$

The transformation between the actual frequency $\omega$ and the normalized frequency $\Omega$ is

$$\Omega = \frac{2}{w}(\omega - \omega_0) = \frac{2}{b}(f - f_0) \quad . \tag{3-2}$$

Thus, the actual attenuation, phase, and delay are

$$A(\omega) = A_L(\Omega) \tag{3-3}$$

$$\varphi(\omega) = \varphi_L(\Omega) \tag{3-4}$$

$$\tau(\omega) = \frac{2}{w}\tau_L(\Omega) \quad . \tag{3-5}$$

The attenuation is usually needed to determine the complexity of the filter.

Three different realizations of three-pole narrow bandpass filters are shown in Figs. 40-42, and the relationship between the design parameters and the circuit element values is given in each case. In Fig. 40 the circuit has three reactively coupled, parallel resonant circuits; in Fig. 41 the circuit has three reactively coupled, series resonant circuits; and in Fig. 42 the circuit is a combination of the previous two. For filters with n poles, n coupled resonant circuits must be used in the realization; the design parameters for an n-pole filter follow the pattern illustrated in the examples for three poles. In all cases the self-admittance of each node and the self-impedance of each loop must be resonant at the center frequency, but whether a coupling reactance is capacitive or inductive is up to the discretion of the designer.

An important fact to observe about these circuits is that the self-admittance of any node is independent of that of the others, as is the self-impedance of the various loops. Thus, the L's and C's can be chosen to be convenient values, e.g., all $L_i$'s in Fig. 40 or in Fig. 41 can be the same value. In Fig. 42, the coupling coefficient $k_{23}$ imposes a relation between $L_2$ and $L_3$, but otherwise in a circuit of this sort the L's are unrestricted.

If a coupling reactance is inductive, the resulting $\pi$ or T of inductors can be converted into a transformer by the equivalent circuits and the equations relating their parameters as shown in Fig. 43. For example, if the $\pi$ of inductors ($L_2$, $L_3$, $L_{23}$) in Fig. 40 or the T inductors ($L_2$, $L_3$, $L_{23}$) in Fig. 41 is converted into a transformer, the coupling coefficient K of that transformer is equal to $\rho k_{23}$.

The envelope of the response of a narrow band filter to a unit impulse or a suddenly applied unit amplitude sine wave of frequency $f_0$ can be found in terms of $h_L(t)$ and $g_L(t)$, respectively. Papoulis (Ref. 14, p. 124) has shown that if a narrow band transfer function is symmetric in amplitude about its center frequency, then the envelope of its unit impulse response is equal to two times the unit impulse response of the equivalent lowpass filter. The narrow band filters considered here have transfer functions whose behavior about the center frequency $\omega_0$ is given by Eqs. (B-17), (B-20), and (B-28). By applying Papoulis' result to these transfer functions it is found that the envelope of the unit impulse response of these filters is

$$h_{env}(t) = \frac{2\Gamma}{wF} h_L\left(\frac{w}{2}t\right) \tag{3-6}$$

for Butterworth, Bessel, and odd-order Chebyshev filters, where

$$F = \begin{cases} \sqrt{C_{11}\, C_{33}} \\ \sqrt{L_{11}\, L_{33}} \\ \sqrt{C_{11}\, L_{33}} \end{cases}$$

for Figs. 40, 41, and 42, respectively. For even-order Chebyshev filters it is

$$h_{env}(t) = \frac{2\Gamma}{wF} \sqrt{1 + \epsilon^2}\, h_L \left(\frac{w}{2} t\right) \quad . \tag{3-7}$$

Similarly, the envelope of the response of a narrow band filter to a suddenly applied sine wave of frequency $f_o$ is

$$g_{env}(t) = \frac{\Gamma}{wF}\, g_L \left(\frac{w}{2} t\right) \tag{3-8}$$

for Butterworth, Bessel, and odd-order Chebyshev filters. For even-order Chebyshev filters it is

$$g_{env}(t) = \frac{\Gamma}{wF} \sqrt{1 + \epsilon^2}\, g_L \left(\frac{w}{2} t\right) \quad . \tag{3-9}$$

## B. LOWPASS FILTERS

For lowpass filters the normalizing factor for the data is just the bandwidth $w = 2\pi b$, and the transformation between $\Omega$ and $\omega$ is

$$\Omega = \omega/w = f/b \quad . \tag{3-10}$$

Thus, the actual attenuation, phase, and delay are

$$A(\omega) = A_L(\Omega) \tag{3-11}$$

$$\varphi(\omega) = \varphi_L(\Omega) \tag{3-12}$$

$$\tau(\omega) = \frac{1}{w} \tau_L(\Omega) \quad . \tag{3-13}$$

Three- and four-pole, polynomial type lowpass filters are shown in Figs. 44 and 45, respectively, and the relationship between the design parameters and the circuit element values is given.* These examples illustrate the definitions of the normalized coupling coefficients and dissipation factors for lowpass filters with either an odd or an even number of poles. Thus for

*Ordinarily, the dissipation factor of inductors is somewhat greater than that of capacitors, so in order to have uniform dissipation, loss (in the form of shunt resistors) must be added to the capacitors. In principle this is undesirable because of the additional transmission loss introduced. However, by incorporating with the source resistance $G_s$ that part of $G_1$ not associated with the parasitic dissipation of $C_1$, the increase in transmission loss is minimized. For n even this can be done at only one end of the network, but for n odd both ends of the network can be treated in this way. Often capacitor losses are negligible, therefore the work of Geffe[15] and Orchard[16] concerned with predistortion for singly loaded, lossy-L networks is of interest. For doubly loaded networks with unequal but uniform dissipation in inductors and capacitors, the analytical work is considerably more difficult (see Desoer,[17] Ming,[18] and Belevitch, et al.[19]), so that in a practical sense the small additional loss required by a design assuming uniform loss in all reactances is well worth the savings in computation. This will no longer be true when a computer program is available for doing this more difficult design.

any value of n and for the dual circuits, the expressions for the design parameters can be found easily from the pattern in these examples.

Unit impulse and unit step responses of the actual lowpass filter are obtained easily from those of the normalized lowpass filter by applying the scaling theorem of Fourier integral theory (Papoulis,[14] p. 14). This theorem states that if $F(\omega)$ and $f(t)$ are a Fourier transform pair, then $F(\omega/a)$ and $|a|\, f(at)$ are also. Thus the unit impulse response of a lowpass filter of bandwidth $w$ is

$$h(t) = \begin{cases} \dfrac{\Gamma}{\sqrt{C_1 C_n}}\, h_L(wt) & \text{for BU, BE and CH with } n \text{ odd.} \\[2ex] \dfrac{\Gamma}{\sqrt{C_1 L_n}}\, h_L(wt) & \text{for BU and BE with } n \text{ even.} \\[2ex] \dfrac{\Gamma\sqrt{1+\epsilon^2}}{\sqrt{C_1 L_n}}\, h_L(wt) & \text{for CH with } n \text{ even.} \end{cases} \tag{3-14}$$

By integrating these expressions we obtain the unit step response of a lowpass filter.

$$g(t) = \begin{cases} \dfrac{\Gamma}{w\sqrt{C_1 C_n}}\, g_L(wt) & \text{for BU, BE and CH with } n \text{ odd.} \\[2ex] \dfrac{\Gamma}{w\sqrt{C_1 L_n}}\, g_L(wt) & \text{for BU and BE with } n \text{ even.} \\[2ex] \dfrac{\Gamma\sqrt{1+\epsilon^2}}{w\sqrt{C_1 L_n}}\, g_L(wt) & \text{for CH with } n \text{ even.} \end{cases} \tag{3-15}$$

Note that for $n$ odd the response is the output voltage, and for $n$ even the response is the output current. In any case the input is a current source.

## C. WIDE BANDPASS FILTERS

For wide bandpass filters the normalizing factor is the bandwidth $w = 2\pi b = \omega_2 - \omega_1 = 2\pi(f_2 - f_1)$, when the pass band extends from $\omega_1 = 2\pi f_1$ to $\omega_2 = 2\pi f_2$. The transformation between $\Omega$ and $\omega$ is

$$\Omega = \frac{\omega_0}{w}\left(\frac{\omega}{\omega_0} - \frac{\omega_0}{\omega}\right) \tag{3-16}$$

where the center frequency $\omega_0 = 2\pi f_0$ is the geometric mean of $\omega_1$ and $\omega_2$: $\omega_0 = \sqrt{\omega_1 \omega_2}$. The actual attenuation, phase, and delay are

$$A(\omega) = A_L(\Omega) \tag{3-17}$$

$$\varphi(\omega) = \varphi_L(\Omega) \tag{3-18}$$

$$\tau(\omega) = \frac{1}{w}\left(1 + \frac{\omega_0^2}{\omega^2}\right)\tau_L(\Omega) \quad . \tag{3-19}$$

A wide bandpass filter of bandwidth w is derived from the equivalent lowpass filter of bandwidth w. The lowpass circuit is modified as follows. In series with each inductance of $L_i$ henrys place a capacitance of $1/\omega_0^2 L_i$ farads, and in parallel with each capacitance of $C_i$ farads place an inductance of $1/\omega_0^2 C_i$ henrys. The resulting L-C pairs are resonant at $\omega_0$. In theory the added reactances must be lossless, but in practice loss in all reactances can be tolerated. If the normalized dissipation factors of the resonant circuits are equal and are smaller than $1/5\ a_{max}$ ($a_{max}$ is the largest value of a in a particular set of curves), the resulting pass-band distortion is negligible for most applications.* The design is summarized in Fig. 46.

In general, the transient response of a wide bandpass filter cannot be related to $h_L(t)$ or $g_L(t)$, so each transient must be handled individually.

## D. SUMMARY

To close the discussion it is worthwhile to outline a general scheme to be used in the design of any of the various filter types and then illustrate by an example.

(1) Determine w or b as discussed in Sec. II.

(2) Convert the specifications to those on an equivalent lowpass filter by normalizing the frequency scale according to Eq. (3-2), (3-10), or (3-16).

(3) Determine n from the required attenuation or other specification.

(4) Make a tentative selection of components and determine the normalized, unloaded dissipation factor, a.

(5) For the appropriate set of curves, choose a value of d which, with the above value of a, determines $\delta$, the k's, and $\Gamma$. To minimize the sensitivity of the filter to errors in the circuit element values, operate where the curves have the least slope.

(6) Compute circuit component values according to the figure that applies to the circuit being designed.

(7) Determine whether the circuit is satisfactory. If not, select new components in step (4) or make any other change that seems feasible and repeat the above steps.

(8) Construct and align according to Appendix A.

The following example illustrates this scheme for the design of a narrow band filter.

Example: A Chebyshev filter with a center frequency of 50 kHz and a ripple bandwidth of 2.0 kHz is required, and the ripple is to be $\gamma = 0.3$ dB. The attenuation should be greater than 40 dB for $f \leqslant 45$ and $f \geqslant 55$ kHz. The circuit is to be that of Fig. 40 with capacitive coupling. According to the above scheme the design proceeds as follows.

(1) Since the ripple bandwidth is specified, $\beta_\alpha = \beta_\gamma = 1$ and b = 2 kHz.

(2) The frequency transformation of Eq. (3-2) is then $\Omega = (f - f_0) \times 10^{-3}$. So the frequencies 45 and 55 kHz correspond to $\Omega = \pm 5$ rad/sec.

(3) We see from the attenuation curves for a CH filter with $\alpha = 0.3$ dB that $A_L(5) = 42$ dB for n = 3. Thus for this filter we have n = 3.

(4) Identical inductors of 0.56 mH are selected. These are made using a type 1811P-A160-3D3 Ferroxcube pot core wound with 59 turns of 61/44 litz wire. The tuning capacitors are to be polystyrene, resulting in $Q_0 = 400$ at 50 kHz. The total resonating capacitance is $C_0 = 18.1$ nF. Since $\rho = b/f_0 = 0.04$, then $a = 1/\rho Q_0 = 0.0625$, which is considerably less than $a_{max}$ for the CH 3-0.3 filter.

*In this case the dissipation in each tuned circuit can be divided arbitrarily between the inductor and the capacitor.

(5) We continue and choose d = 0.5. Then from the curves we read the values:

$$\delta = 0.897$$
$$k_{12} = 0.822$$
$$k_{23} = 0.775$$
$$\Gamma = 0.680 \quad .$$

(6) Our circuit is shown in Fig. 47, and the circuit element values are computed according to Fig. 40.

$$C_{12} = \rho C_0 k_{12} = 595\,\text{pF}$$
$$C_{23} = \rho C_0 k_{23} = 561\,\text{pF}$$
$$C_1 = C_0 - C_{12} \approx 17.5\,\text{nF}$$
$$C_2 = C_0 - C_{12} - C_{23} \approx 17.0\,\text{nF}$$
$$C_3 = C_0 - C_{23} \approx 17.5\,\text{nF}$$
$$R_s = \frac{1}{2\pi b C_0 (d - a)} = 8.19\,\text{k}\Omega$$
$$R_\ell = \frac{1}{2\pi b C_0 (\delta - a)} = 5.27\,\text{k}\Omega \quad .$$

The transfer impedance at 50 kHz is

$$\left|\frac{E_3}{I_1}\right| = \frac{\Gamma}{2\pi b C_0} = 3.01\,\text{k}\Omega \quad .$$

(7) The component values are satisfactory so the design is complete.

The filter was constructed and aligned according to the method outlined in Appendix A. Its transfer impedance at 50 kHz was measured as 2.86 kΩ which is a 5-percent error compared to the theoretical value of 3.01 kΩ. Its measured amplitude vs frequency response is shown in Fig. 48. The pass band is quite close to the designed behavior, but the skirts are a bit unsymmetrical. The attenuation on the low frequency side of the pass band increases at a slightly higher rate than that on the high side because of the zeros of transmission at zero frequency and the cluster of three poles in the vicinity of −50 kHz. Since the circuit has capacitive coupling, there are five transmission zeros at $\omega = 0$. If we approximate the effect of the poles by assuming that they are concentrated at −50 kHz, then the overall distortion caused by these three poles and five zeros relative to that at 50 kHz is given by the factor $\left(\frac{2}{1 + f/50}\right)^3 \left(\frac{f}{50}\right)^5$, where f is in kilohertz. If the effect of this factor is removed from the measured curve in Fig. 48, the pass band is essentially unchanged, but the skirts are shifted to pass through the squares. The shifted curve (which is the response of the three poles in the vicinity of 50 kHz) is nearly symmetric* and is in close agreement with the theoretical curve, $A_L(f - 50)$, (17 dB at 48 and 52 kHz, 42 dB at 45 and 55 kHz, and 60.5 dB at 40 and 60 kHz). Thus, the amount of skewing can be predicted quite accurately. Note that skewing can be controlled to some extent by changing the kind of coupling reactances that are used, thereby changing the number of transmission zeros at zero frequency. In the above circuit, for example, if one coupling reactance were capacitive and the other inductive, there would be three transmission zeros at zero frequency. The skewness would be less than 1 dB at 40 and 60 kHz.

---

*Within the accuracy of the narrow band design, the poles are placed symmetrically about the line $s = \sigma + i\,\omega_0$.

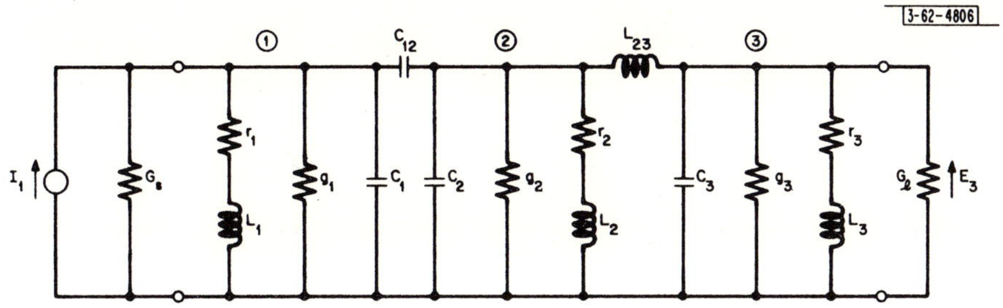


Normalized coupling coefficients ($\rho = w/\omega_0 = b/f_0$):

$$k_{12} = \frac{C_{12}}{\rho\sqrt{C_{11}\,C_{22}}} \quad \text{and} \quad k_{23} = \frac{\sqrt{L_{22}L_{33}}}{\rho L_{23}}$$

where

$$C_{11} = \text{self-capacitance of node } 1 = C_1 + C_{12}$$

$$C_{22} = \text{self-capacitance of node } 2 = C_2 + C_{12}$$

$$L_{22} = \text{self-inductance of node } 2 = \frac{L_2\,L_{23}}{L_2 + L_{23}}$$

$$L_{33} = \text{self-inductance of node } 3 = \frac{L_3\,L_{23}}{L_3 + L_{23}} \quad .$$

Normalized parasitic dissipation factors:

$$a = \frac{D_0}{\rho} = \frac{1}{\rho Q_0} = \frac{1}{\rho}\left(\frac{r_i}{\omega_0 L_i} + \frac{g_i}{\omega_0 C_i}\right) \quad \text{for } i = 1,2, \text{ and } 3$$

where $D_0 = Q_0^{-1}$ is the parasitic dissipation factor of a tuned circuit.

Source and load conductances:

$$G_s = wC_{11}(d - a) \quad \text{and} \quad G_\ell = wC_{33}(\delta - a) \quad .$$

Transfer impedance at $\omega_0$:

$$\left|\frac{E_3}{I_1}\right|_{\omega = \omega_0} = \frac{\Gamma}{w\sqrt{C_{11}C_{33}}}$$

Fig. 40. Network (using coupled parallel resonant circuits) and equations for realization of three-pole narrow bandpass filter.

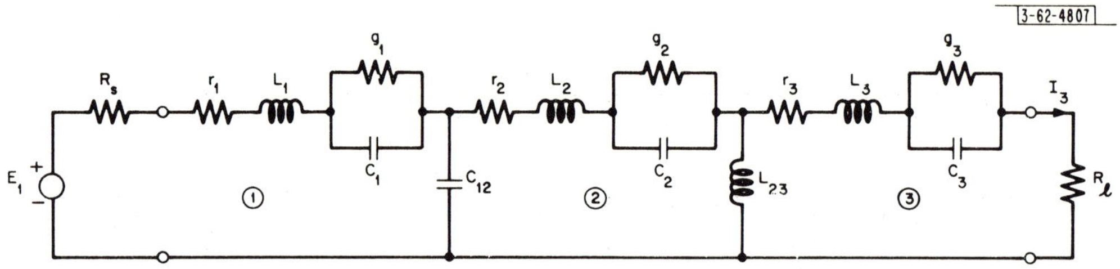


Normalized coupling coefficients ($\rho = w/\omega_0 = b/f_0$):

$$k_{12} = \frac{\sqrt{C_{11}\,C_{22}}}{\rho C_{12}} \quad \text{and} \quad k_{23} = \frac{L_{23}}{\rho\sqrt{L_{23}L_{23}}}$$

where

$$C_{11} = \text{self-capacitance of loop 1} = \frac{C_1 C_{12}}{C_1 + C_{12}}$$

$$C_{22} = \text{self-capacitance of loop 2} = \frac{C_2 C_{12}}{C_2 + C_{12}}$$

$$L_{22} = \text{self-inductance of loop 2} = L_2 + L_{23}$$

$$L_{33} = \text{self-inductance of loop 3} = L_3 + L_{23}$$

Normalized parasitic dissipation factors:

$$a = \frac{D_0}{\rho} = \frac{1}{\rho Q_0} = \frac{1}{\rho}\left(\frac{r_i}{\omega_0 L_i} + \frac{g_i}{\omega_0 C_i}\right) \qquad \text{for } i = 1, 2, \text{ and } 3$$

where $D_0 = Q_0^{-1}$ is the parasitic dissipation factor of a tuned circuit.

Source and load resistance:

$$R_s = wL_{11}(d - a) \quad \text{and} \quad R_\ell = wL_{33}(\delta - a)$$

Transfer admittance at $\omega_0$:

$$\left|\frac{I_3}{E_1}\right|_{\omega = \omega_0} = \frac{\Gamma}{w\sqrt{L_{11}L_{33}}}$$

Fig. 41. Network (using coupled series resonant circuits) and equations for realization of three-pole narrow bandpass fil.er.

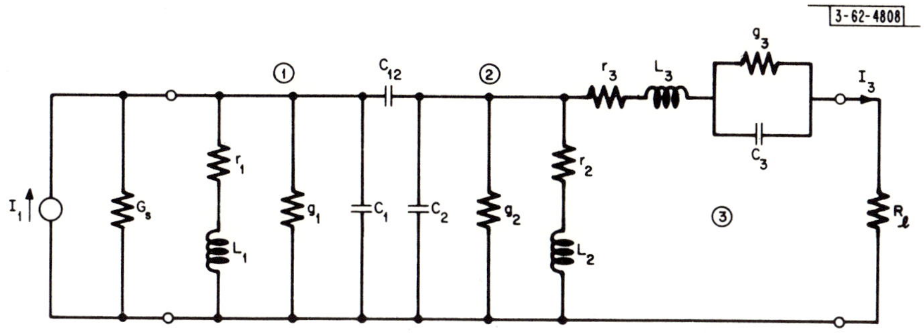


Normalized coupling coefficients ($\rho = w/\omega_0 = b/f_0$):

$$k_{12} = \frac{C_{12}}{\rho\sqrt{C_{11}\,C_{22}}} \quad \text{and} \quad k_{23} = \frac{1}{w\sqrt{C_{22}L_{33}}} = \frac{1}{\rho}\sqrt{\frac{L_{22}}{L_{33}}}$$

where

$C_{11}$ = self-capacitance of node 1 = $C_1 + C_{12}$

$C_{22}$ = self-capacitance of node 2 with loop 3 open = $C_2 + C_{12}$

$L_{22}$ = self-inductance of node 2 with loop 3 open = $L_2$

$L_{33}$ = self-inductance of loop 3 with node 2 shorted = $L_3$ .

Normalized parasitic dissipation factors:

$$a = \frac{D_0}{\rho} = \frac{1}{\rho Q_0} = \frac{1}{\rho}\left(\frac{r_i}{\omega_0 L_i} + \frac{g_i}{\omega_0 C_i}\right) \quad \text{for } i = 1, 2, \text{ and } 3$$

Source conductance and load resistance:

$$G_s = wC_{11}(d - a) \quad \text{and} \quad R_\ell = wL_{33}(\delta - a) \quad .$$

Current transfer ratio at $\omega_0$:

$$\left|\frac{I_3}{I_1}\right|_{\omega=\omega_0} = \frac{\Gamma}{w\sqrt{C_{11}\,L_{33}}} \quad .$$

Fig. 42. Network (using coupled parallel and series resonant circuits) and equations for realization of three-pole narrow bandpass filter.

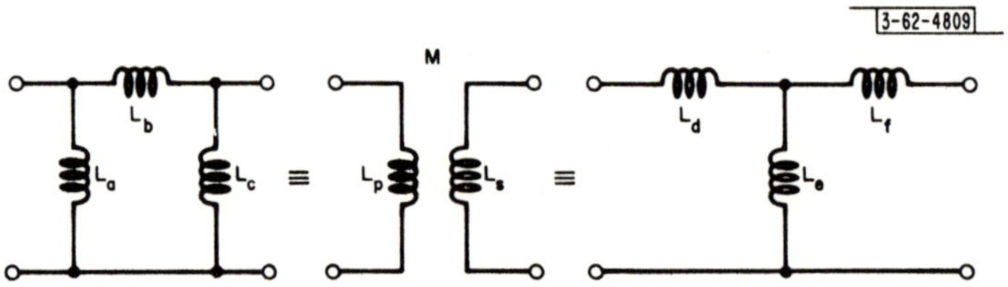


$$L_p = \frac{L_a (L_b + L_c)}{L_a + L_b + L_c} \qquad L_p = L_d + L_e$$

$$L_s = \frac{L_c (L_b + L_a)}{L_a + L_b + L_c} \qquad L_s = L_f + L_e$$

$$|M| = \frac{L_a L_c}{L_a + L_b + L_c} \qquad |M| = L_e$$

$$K = \frac{|M|}{\sqrt{L_p L_s}} = \left[\frac{L_a L_c}{(L_a + L_b)(L_c + L_b)}\right]^{1/2} \qquad K = \frac{|M|}{\sqrt{L_p L_s}} = \frac{L_e}{\sqrt{(L_d + L_e)(L_f + L_e)}}$$

Fig. 43. Equivalent circuits for conversion of a T or π of inductors into a transformer.

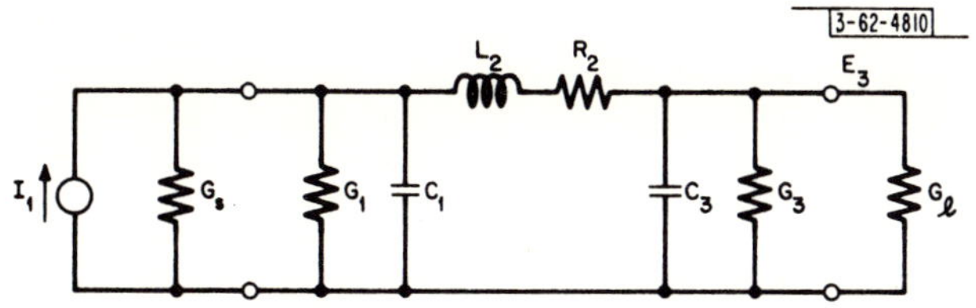


Normalized coupling coefficients:

$$k_{12} = \frac{1}{w\sqrt{C_1 L_2}}$$

$$k_{23} = \frac{1}{w\sqrt{L_2 C_3}} \quad .$$

Normalized dissipation factors:

$$a = \frac{G_1}{wC_1} = \frac{R_2}{wL_2} = \frac{G_3}{wC_3} \quad .$$

Source and load conductances:

$$G_s = wC_1\ (d - a)$$

$$G_\ell = wC_3\ (\delta - a) \quad .$$

Transfer impedance at DC:

$$\left.\frac{E_3}{I_1}\right|_{\omega=0} = \frac{\Gamma}{w\sqrt{C_1 C_3}} \quad .$$

Fig. 44. Three-pole, polynomial type lowpass filter.

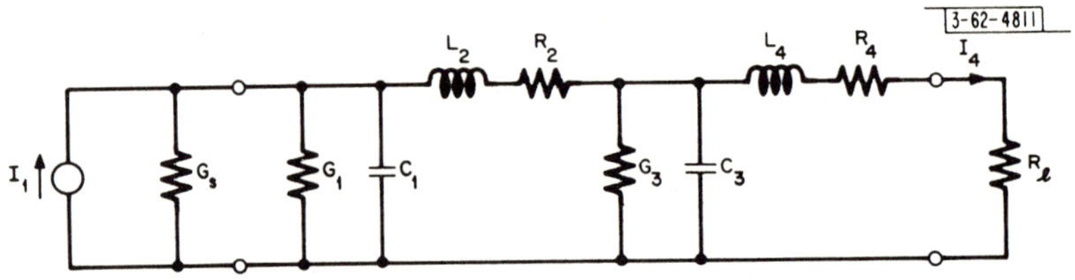


Normalized coupling coefficients:

$$k_{12} = \frac{1}{w\sqrt{C_1 L_2}}$$

$$k_{23} = \frac{1}{w\sqrt{L_2 C_3}}$$

$$k_{34} = \frac{1}{w\sqrt{C_3 L_4}} \quad .$$

Normalized parasitic dissipation factors:

$$a = \frac{G_1}{wC_1} = \frac{R_2}{wL_2} = \frac{G_3}{wC_3} = \frac{R_4}{wL_4} \quad .$$

Source conductance and load resistance:

$$G_s = wC_1 \,(d - a)$$

$$R_\ell = wL_4 \,(\delta - a) \quad .$$

Current transfer ratio at DC:

$$\left.\frac{I_4}{I_1}\right|_{\omega=0} = \frac{\Gamma}{w\sqrt{C_1 L_4}} \quad .$$

Fig. 45. Four-pole, polynomial type lowpass filter.

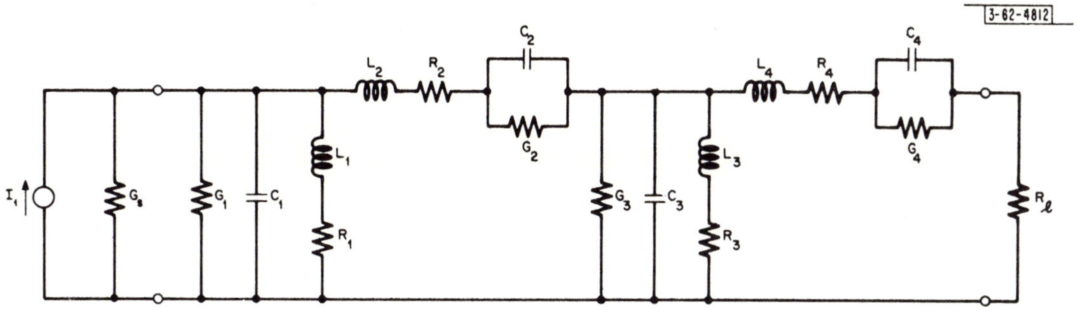


Normalized coupling coefficients:

$$k_{12} = \frac{1}{w\sqrt{C_1 L_2}}$$

$$k_{23} = \frac{1}{w\sqrt{L_2 C_3}}$$

$$k_{34} = \frac{1}{w\sqrt{C_3 L_4}}$$

Tuning:

$$\omega_0^{-2} = L_1 C_1 = L_2 C_2 = L_3 C_3 = L_4 C_4 \quad .$$

Normalized parasitic dissipation factors:

$$a = \frac{1}{\rho}\left(\frac{R_i}{\omega_0 L_i} + \frac{G_i}{\omega_0 C_i}\right) = \frac{1}{\rho}\left(D_{L_i} + D_{C_i}\right) \quad \text{for } i = 1,2,3, \text{ and } 4.$$

Source conductance and load resistance:

$$G_s = wC_1\,(d - a) \quad \text{and} \quad R_\ell = wL_4\,(\delta - a) \quad .$$

Current transfer ratio at $\omega_0$:

$$\left|\frac{I_4}{I_1}\right|_{\omega=\omega_0} = \frac{\Gamma}{w\sqrt{C_1 L_4}} \quad .$$

Fig. 46. Wide bandpass filter derived from four-pole lowpass filter of Fig.6.

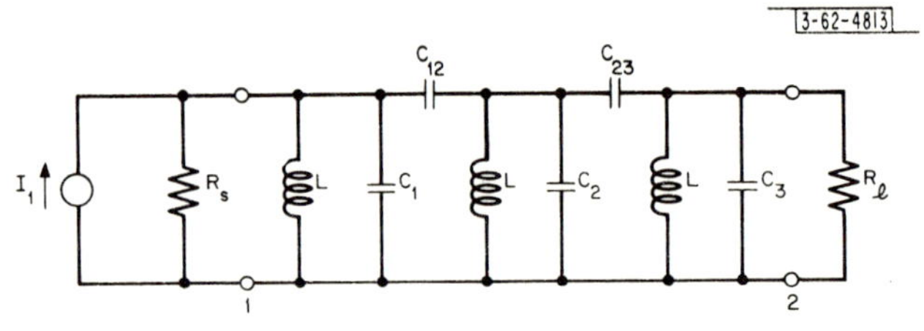


Fig. 47. Circuit for three-pole capacitively coupled bandpass filter.

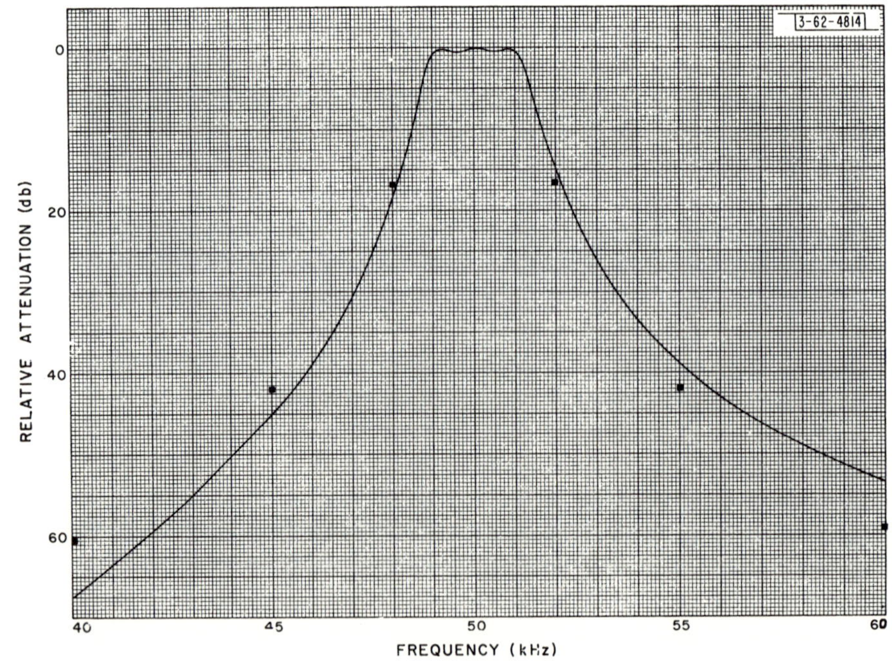


Fig. 48. Measured amplitude vs frequency characteristic of CH 3-0.3 filter in the example.

## IV. DESIGN DATA

In this section we present the design data in Figs. 49-167 in order of increasing filter complexity (n = 2 through 5 poles). For each value of n the data are arranged according to type of transfer function in the order Butterworth, Chebyshev, and Bessel. For Chebyshev filters the data is given in order of increasing passband ripple.

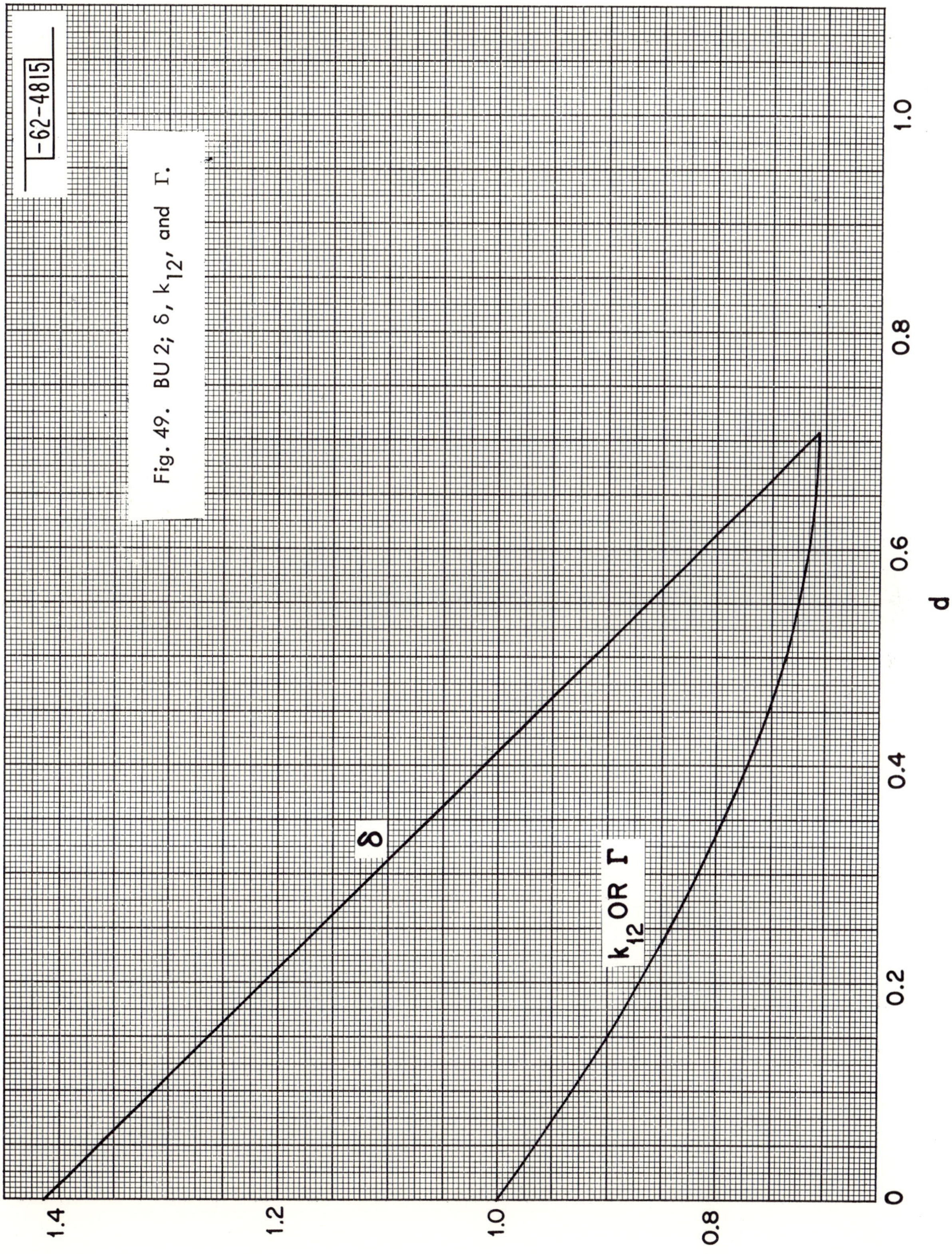


Fig. 49. BU 2; δ, $k_{12}$, and Γ.

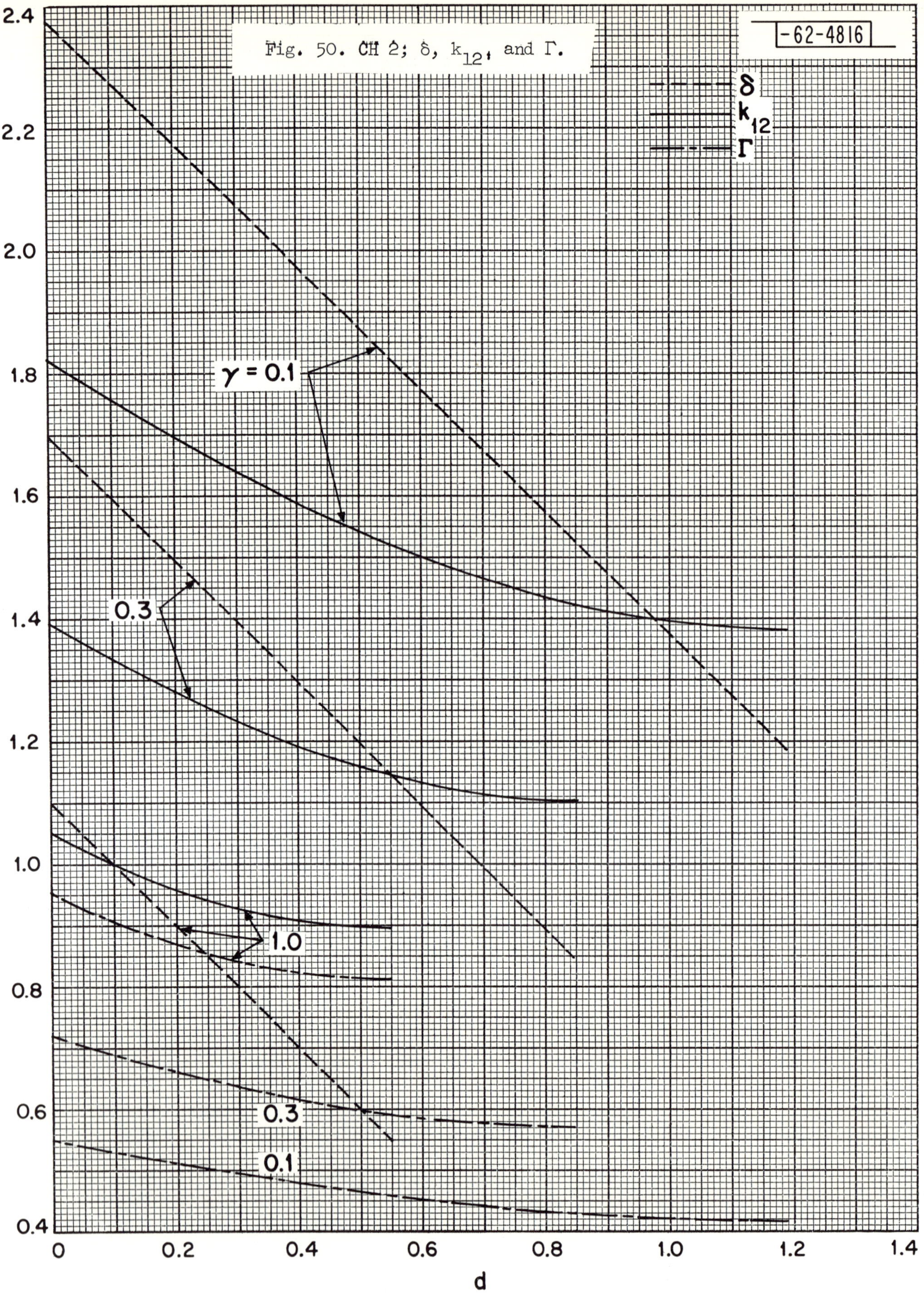
Fig. 50. CH 2; δ, k12, and Γ.
-62-4816
δ
k12
Γ
γ = 0.1
0.3
1.0
0.3
0.1
2.4
2.2
2.0
1.8
1.6
1.4
1.2
1.0
0.8
0.6
0.4
0
0.2
0.4
0.6
0.8
1.0
1.2
1.4
d

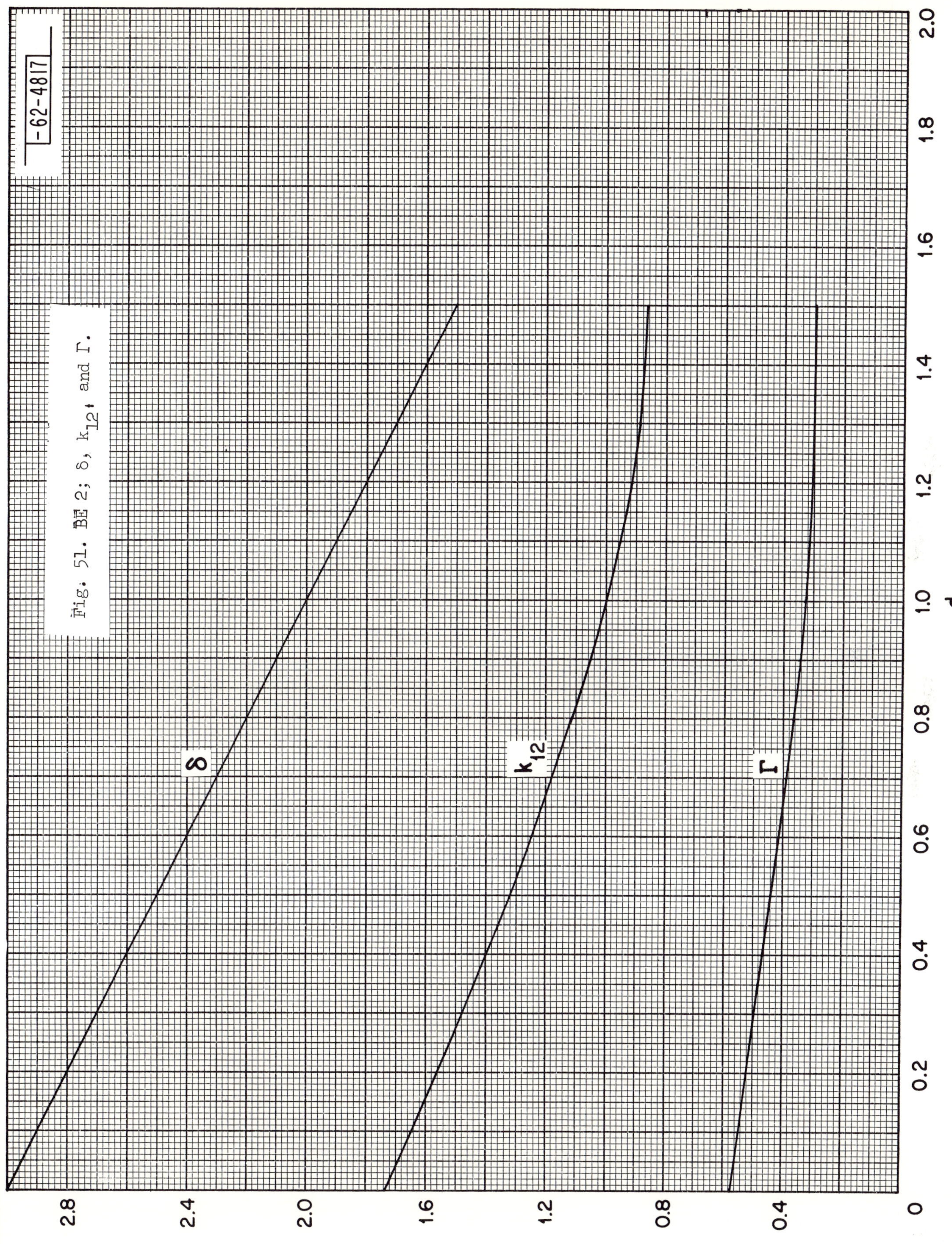


Fig. 51. BE 2; δ, $k_{12}$, and Γ.

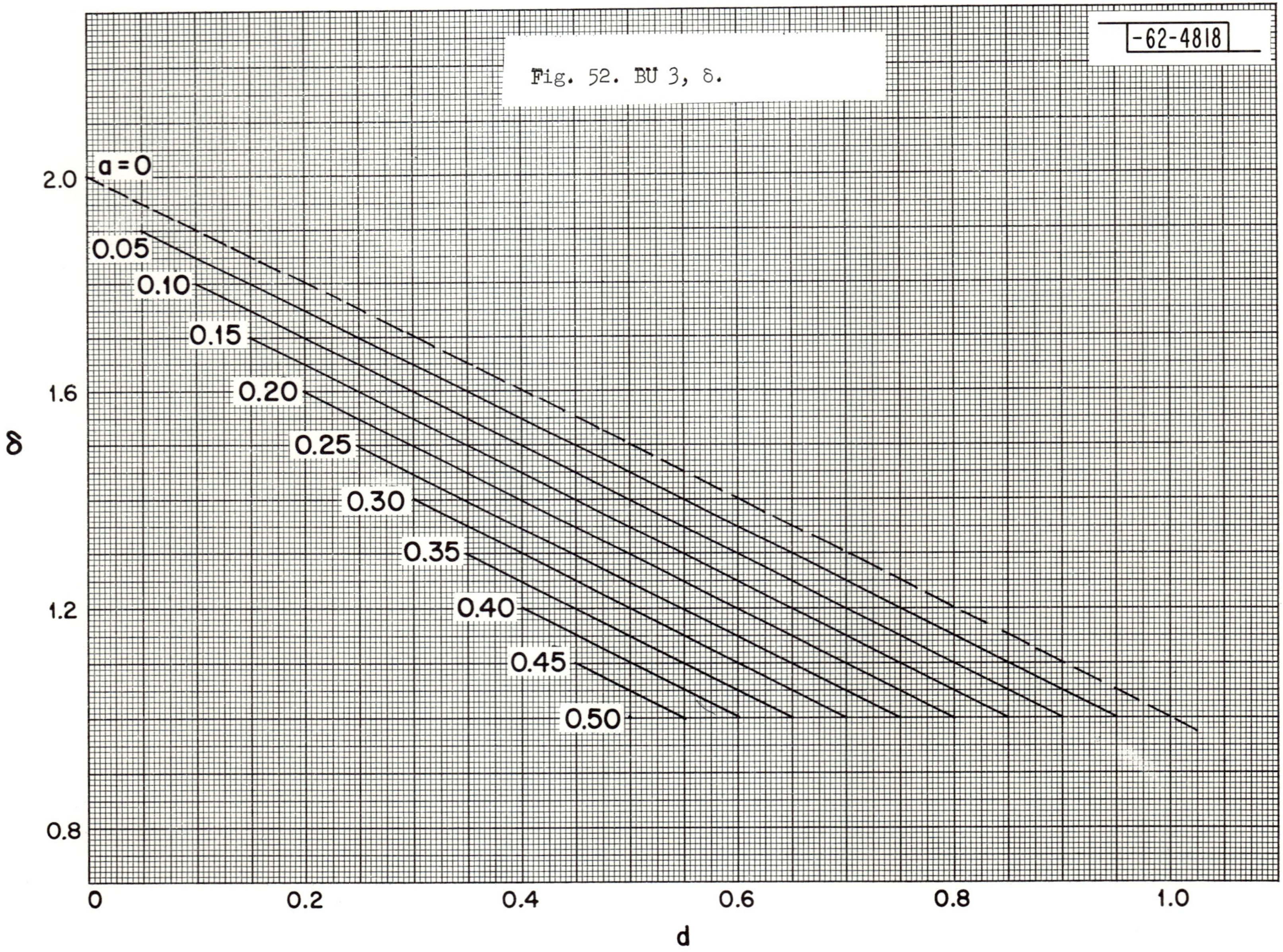
-62-4818
a = 0
0.05
0.10
0.15
0.20
0.25
0.30
0.35
0.40
0.45
0.50
2.0
1.6
1.2
0.8
δ
0
0.2
0.4
0.6
0.8
1.0
d

Fig. 52. BU 3, δ.

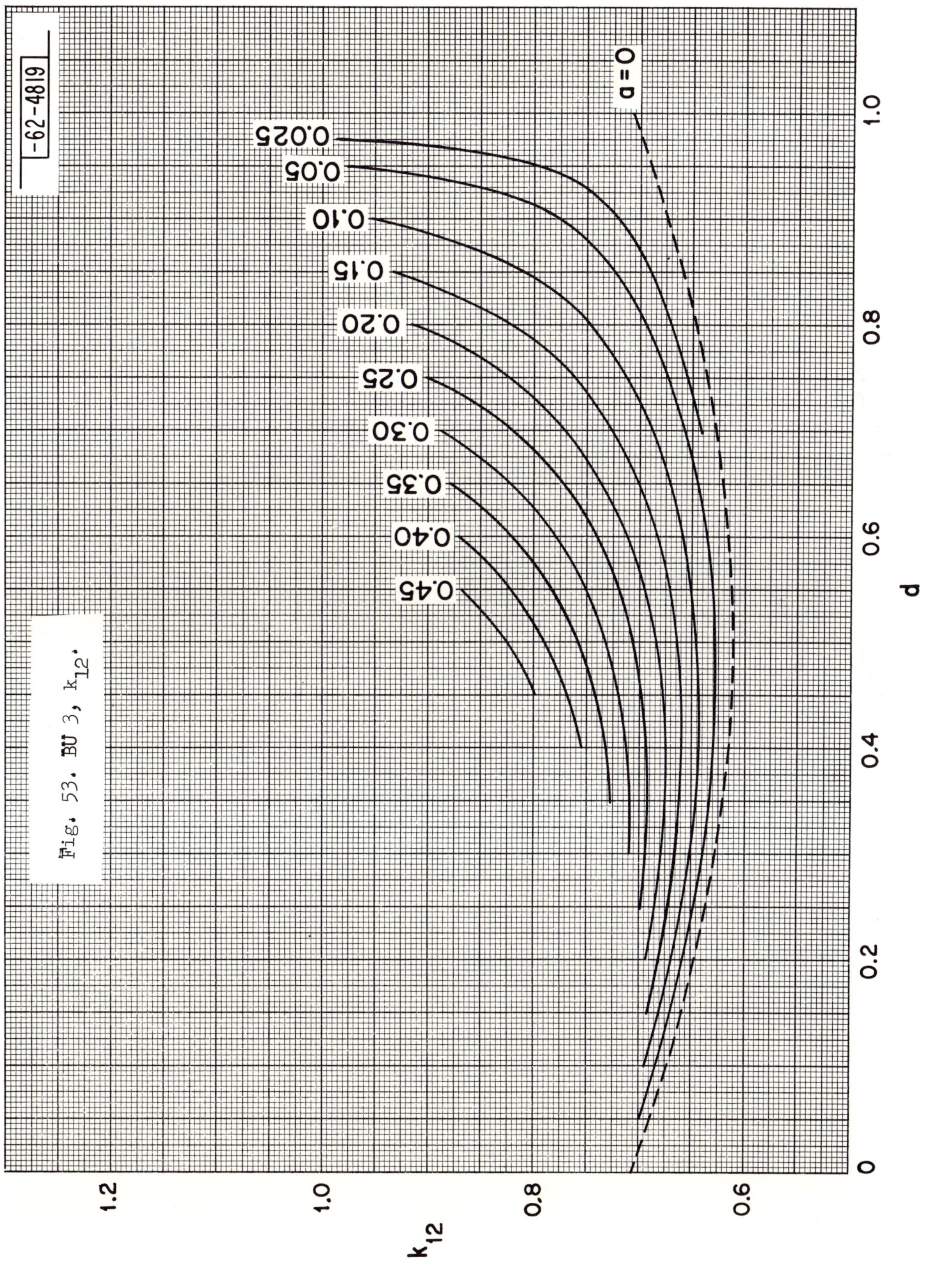


Fig. 53. BU 3, $k_{12}$.

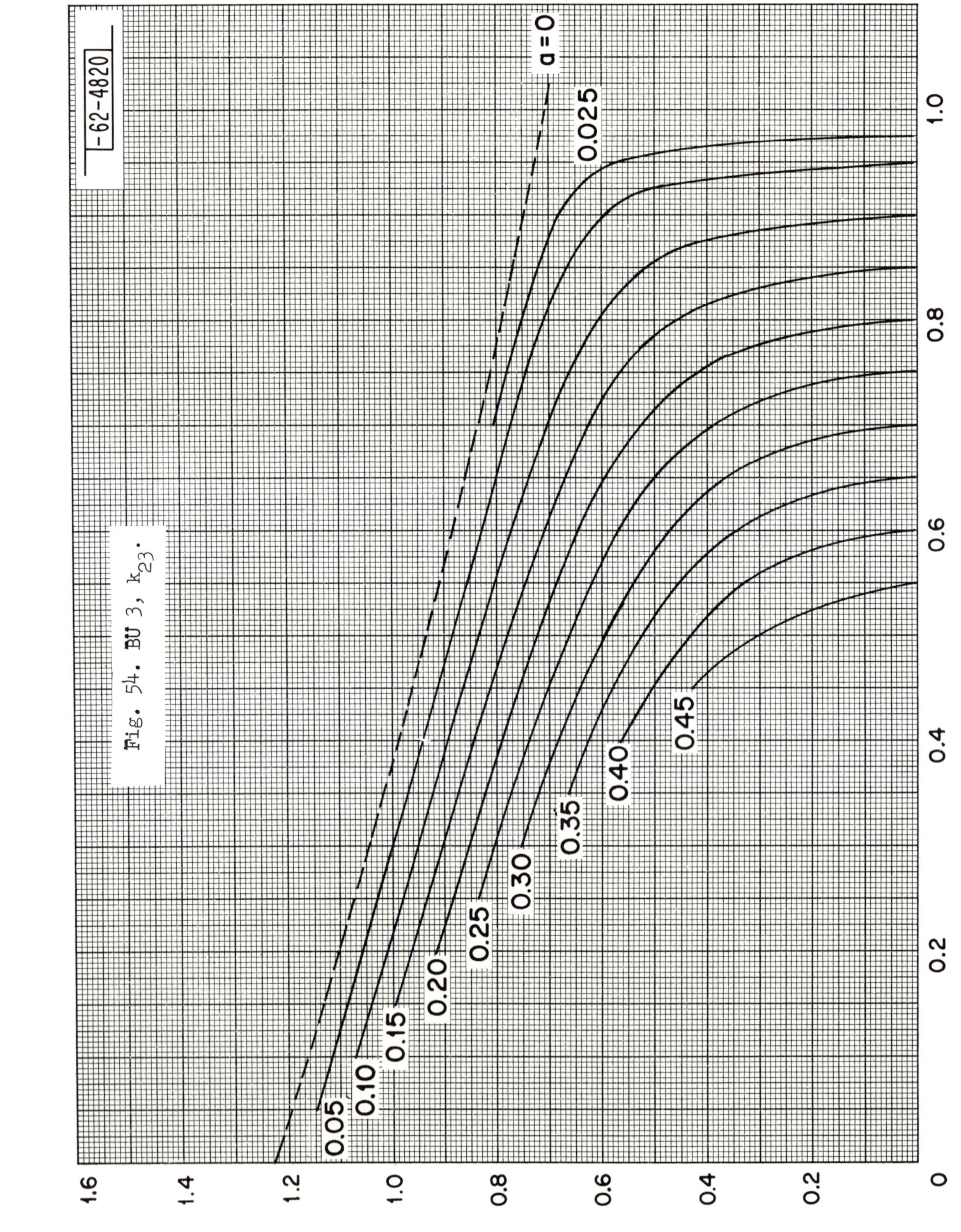


Fig. 54. BU 3, $k_{23}$.

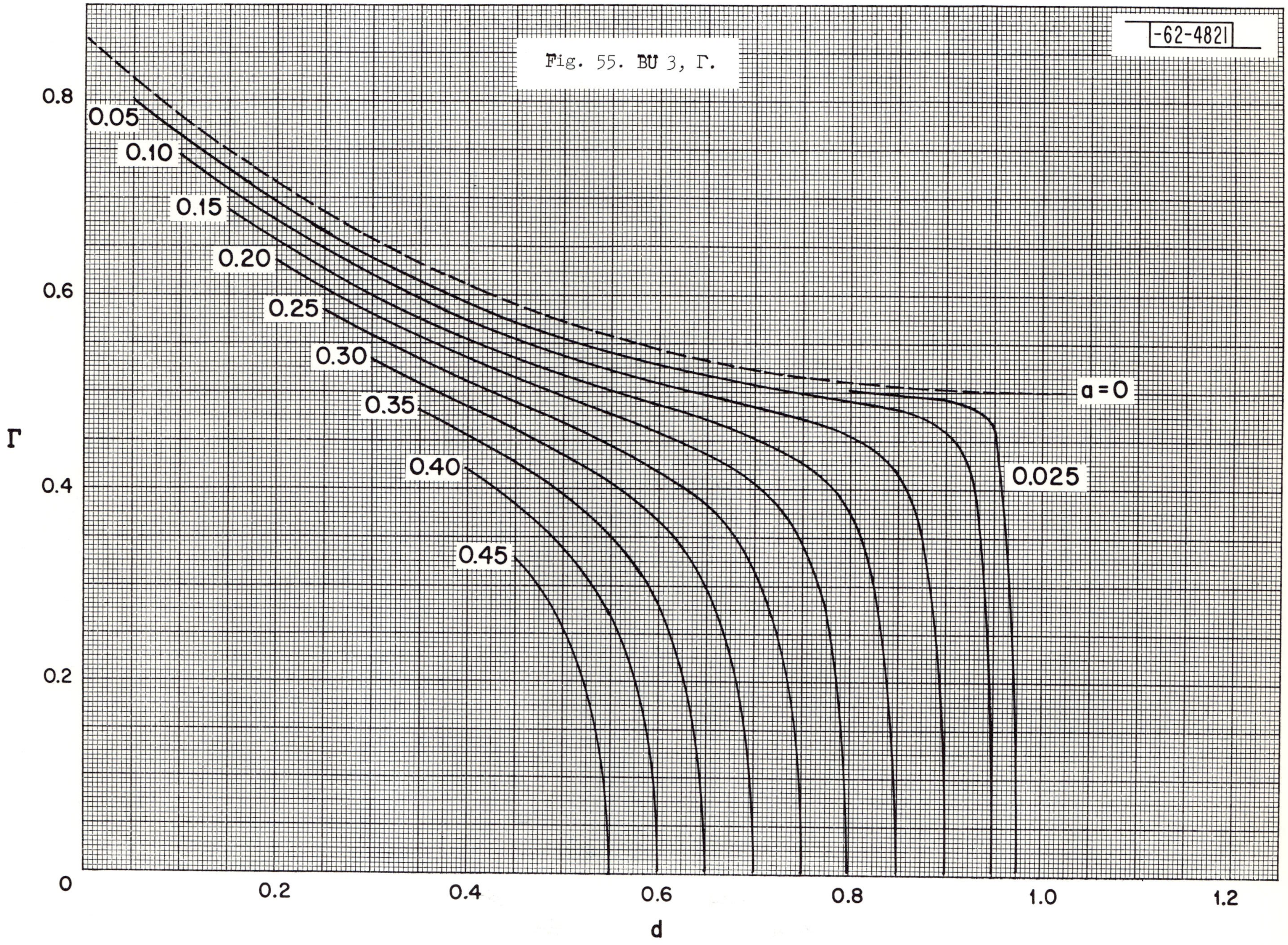
-62-4821
a=0
0.025
0.05
0.10
0.15
0.20
0.25
0.30
0.35
0.40
0.45
Γ
0
0.2
0.4
0.6
0.8
0.2
0.4
0.6
0.8
1.0
1.2
d

Fig. 55. BU 3, Γ.

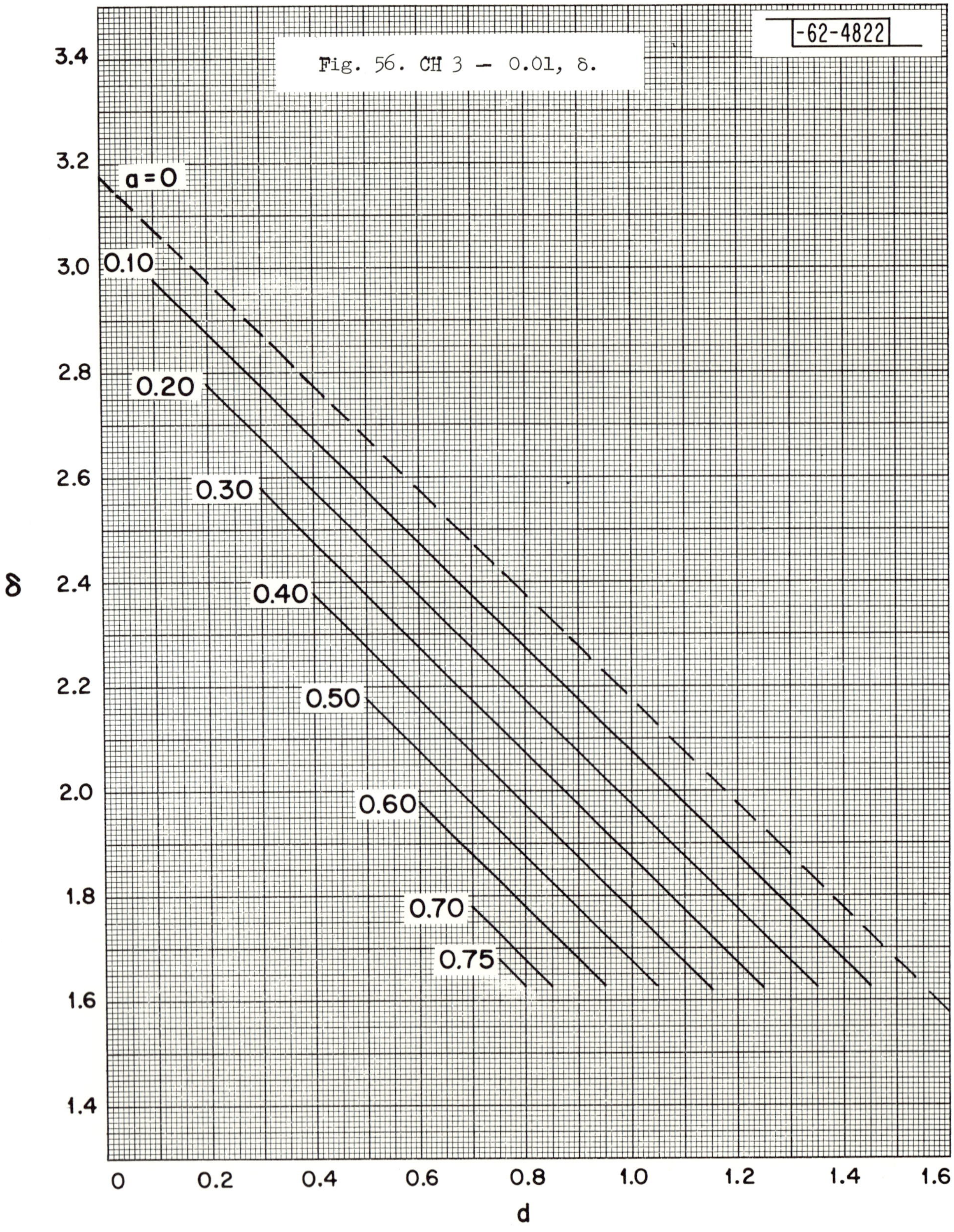


Fig. 56. CH 3 — 0.01, δ.

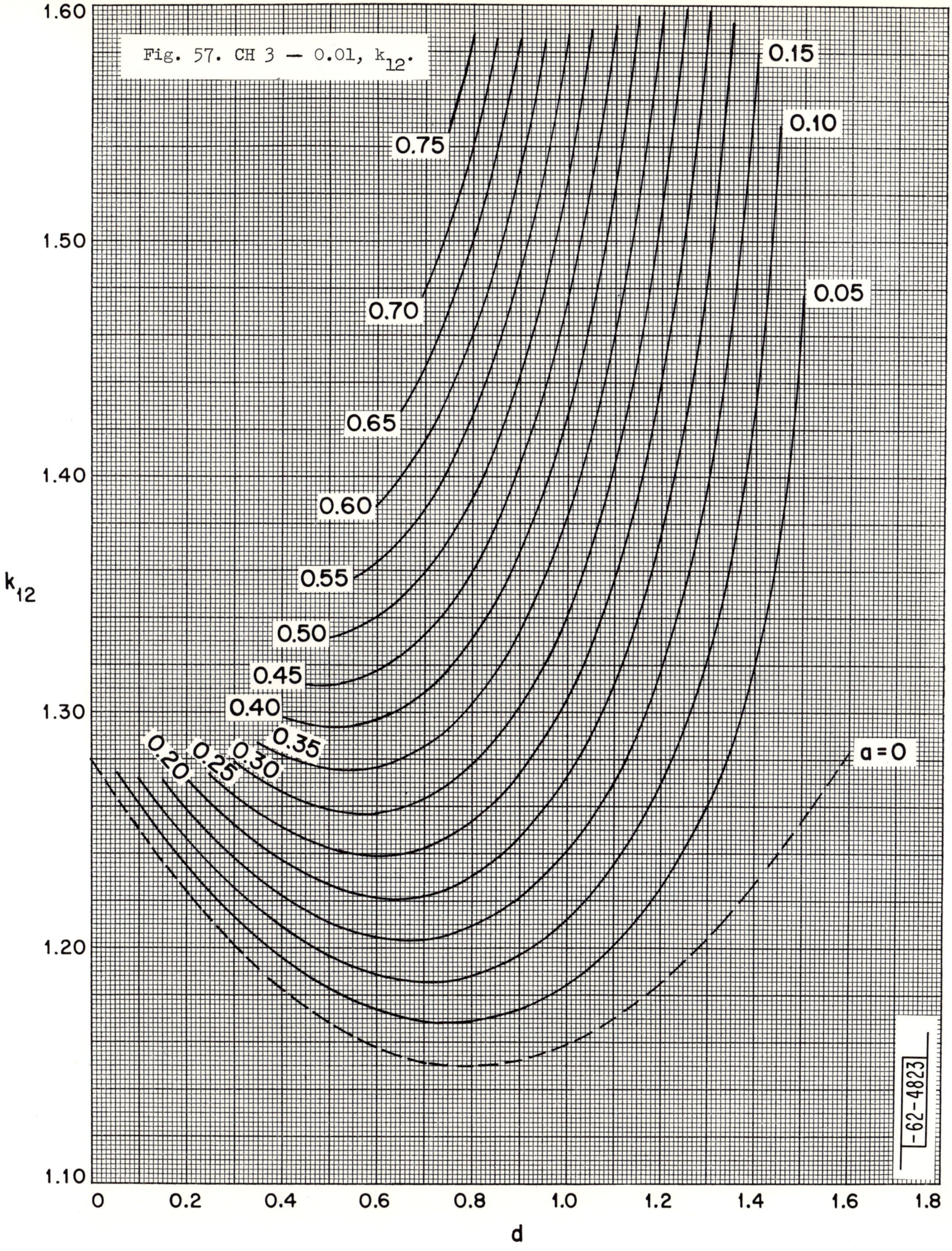
Fig. 57. CH 3 — 0.01, k12.
1.60
1.50
1.40
1.30
1.20
1.10
k12
0
0.2
0.4
0.6
0.8
1.0
1.2
1.4
1.6
1.8
d
0.75
0.70
0.65
0.60
0.55
0.50
0.45
0.40
0.35
0.30
0.25
0.20
0.15
0.10
0.05
a=0
-62-4823

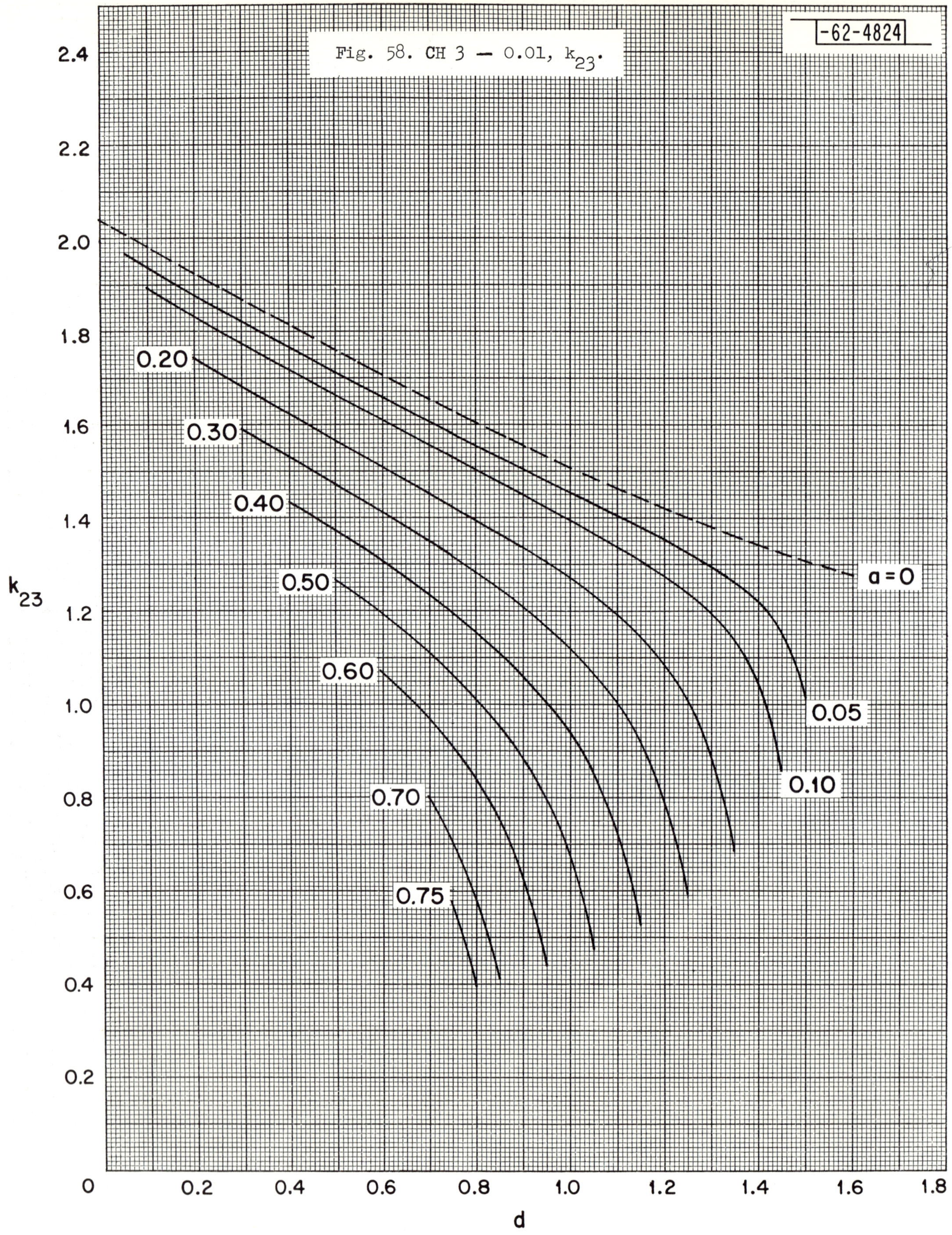
-62-4824
Fig. 58. CH 3 — 0.01, $k_{23}$.
2.4
2.2
2.0
1.8
1.6
1.4
1.2
1.0
0.8
0.6
0.4
0.2
0
$k_{23}$
0.2
0.4
0.6
0.8
1.0
1.2
1.4
1.6
1.8
d
0.20
0.30
0.40
0.50
0.60
0.70
0.75
a = 0
0.05
0.10

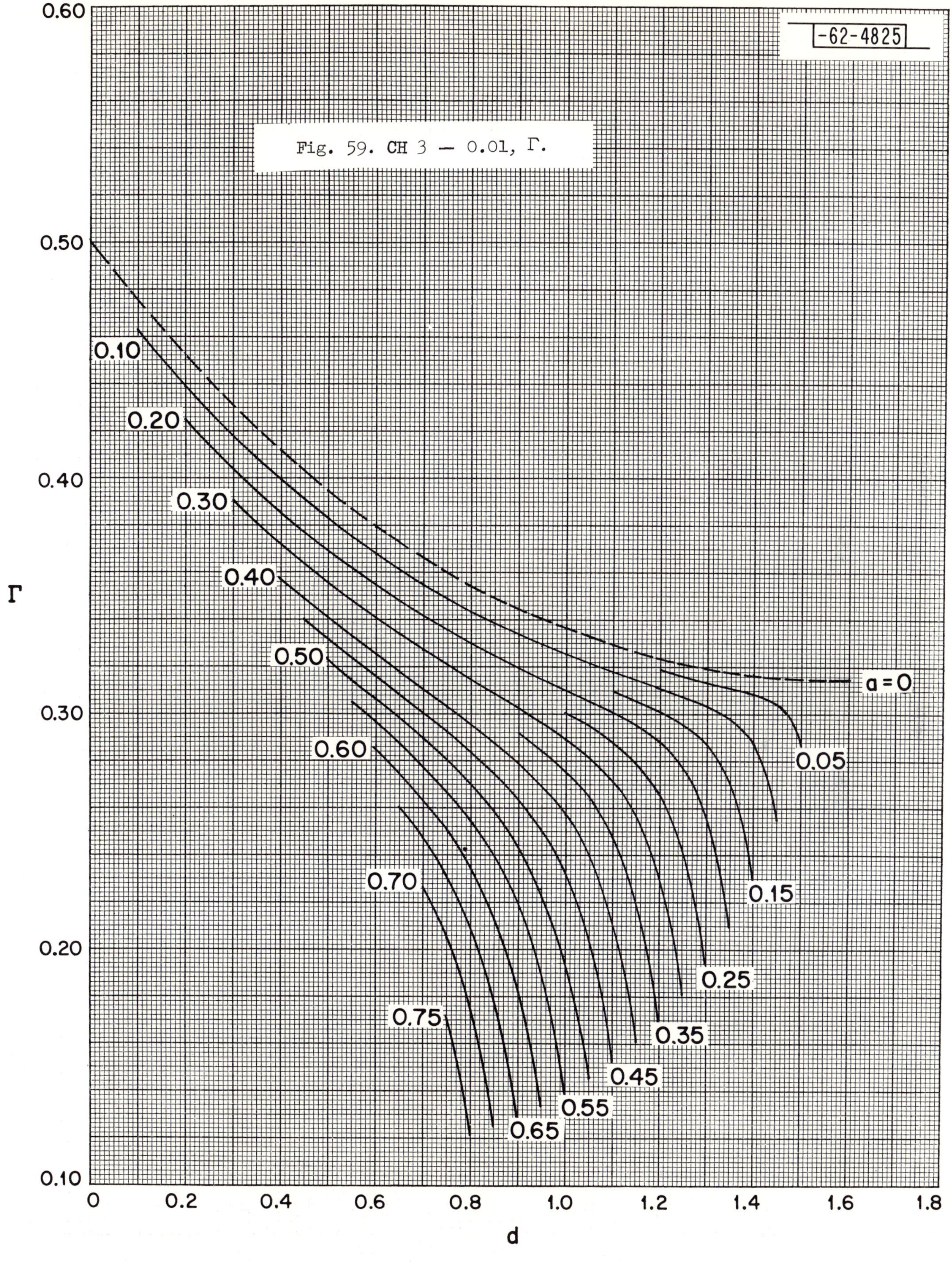
-62-4825
Fig. 59. CH 3 — 0.01, Γ.
0.60
0.50
0.40
0.30
0.20
0.10
Γ
0
0.2
0.4
0.6
0.8
1.0
1.2
1.4
1.6
1.8
d
0.10
0.20
0.30
0.40
0.50
0.60
0.70
0.75
a = 0
0.05
0.15
0.25
0.35
0.45
0.55
0.65

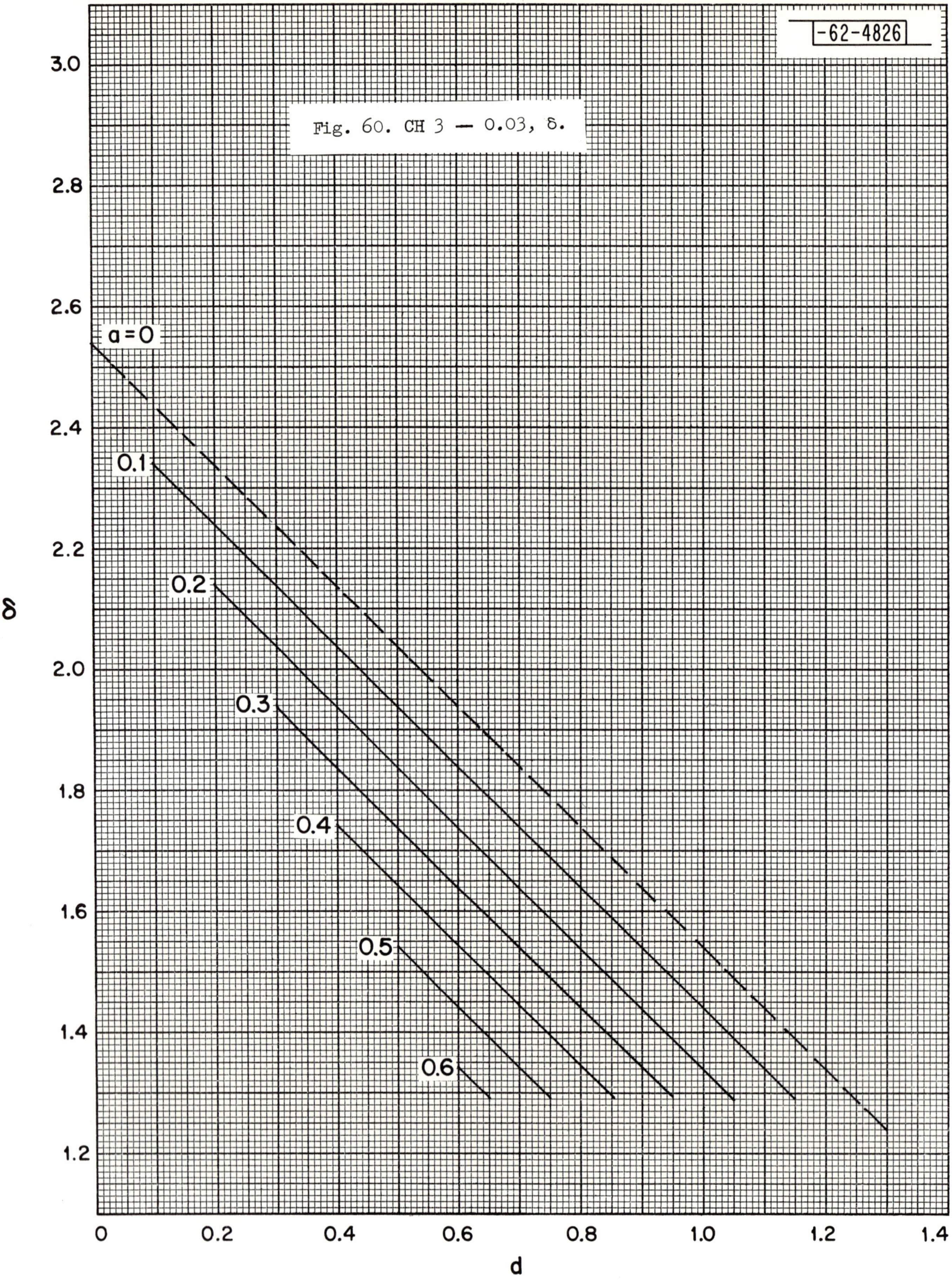
-62-4826
Fig. 60. CH 3 — 0.03, δ.
a=0
0.1
0.2
0.3
0.4
0.5
0.6
δ
3.0
2.8
2.6
2.4
2.2
2.0
1.8
1.6
1.4
1.2
0
0.2
0.4
0.6
0.8
1.0
1.2
1.4
d

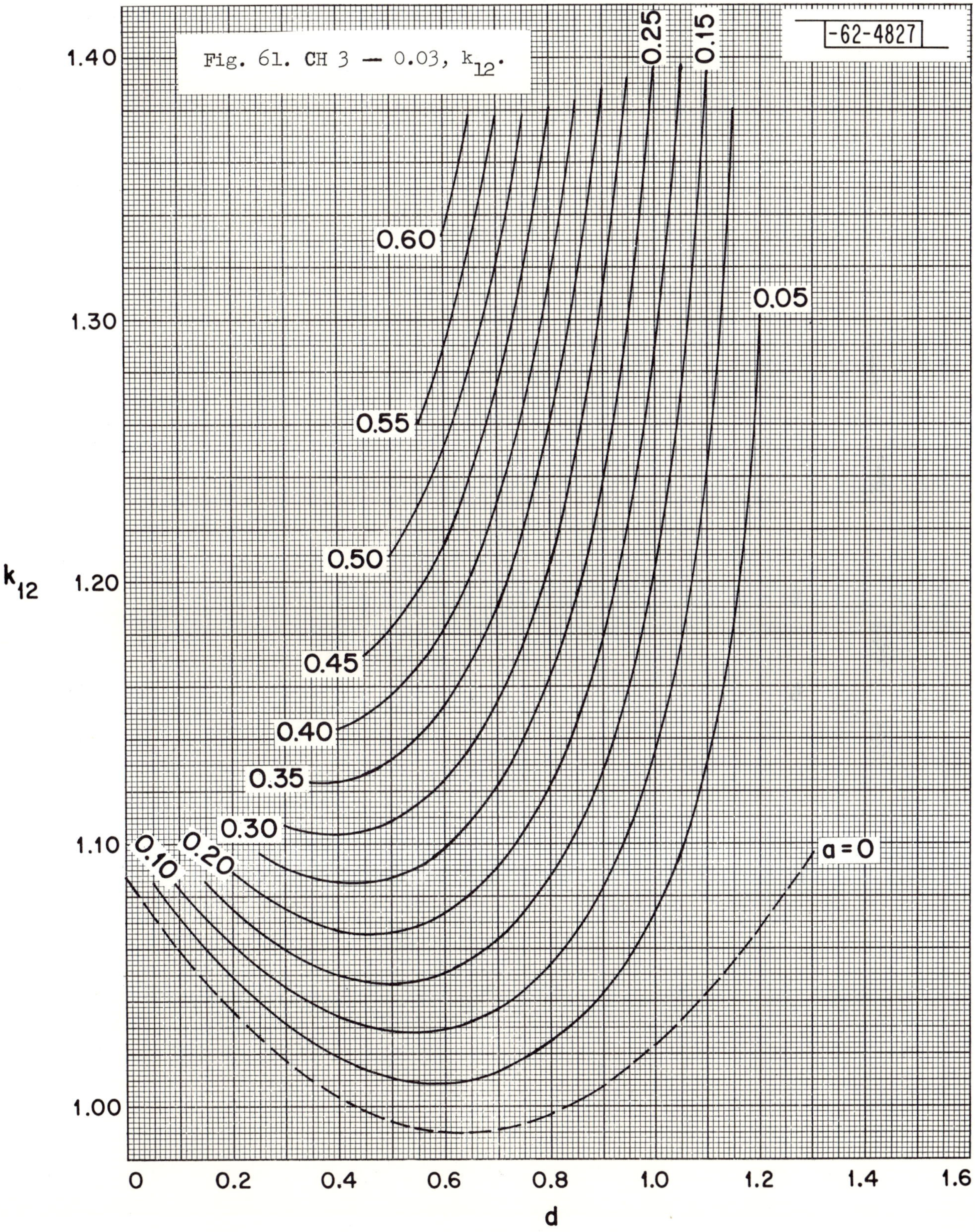


Fig. 61. CH 3 — 0.03, $k_{12}$.

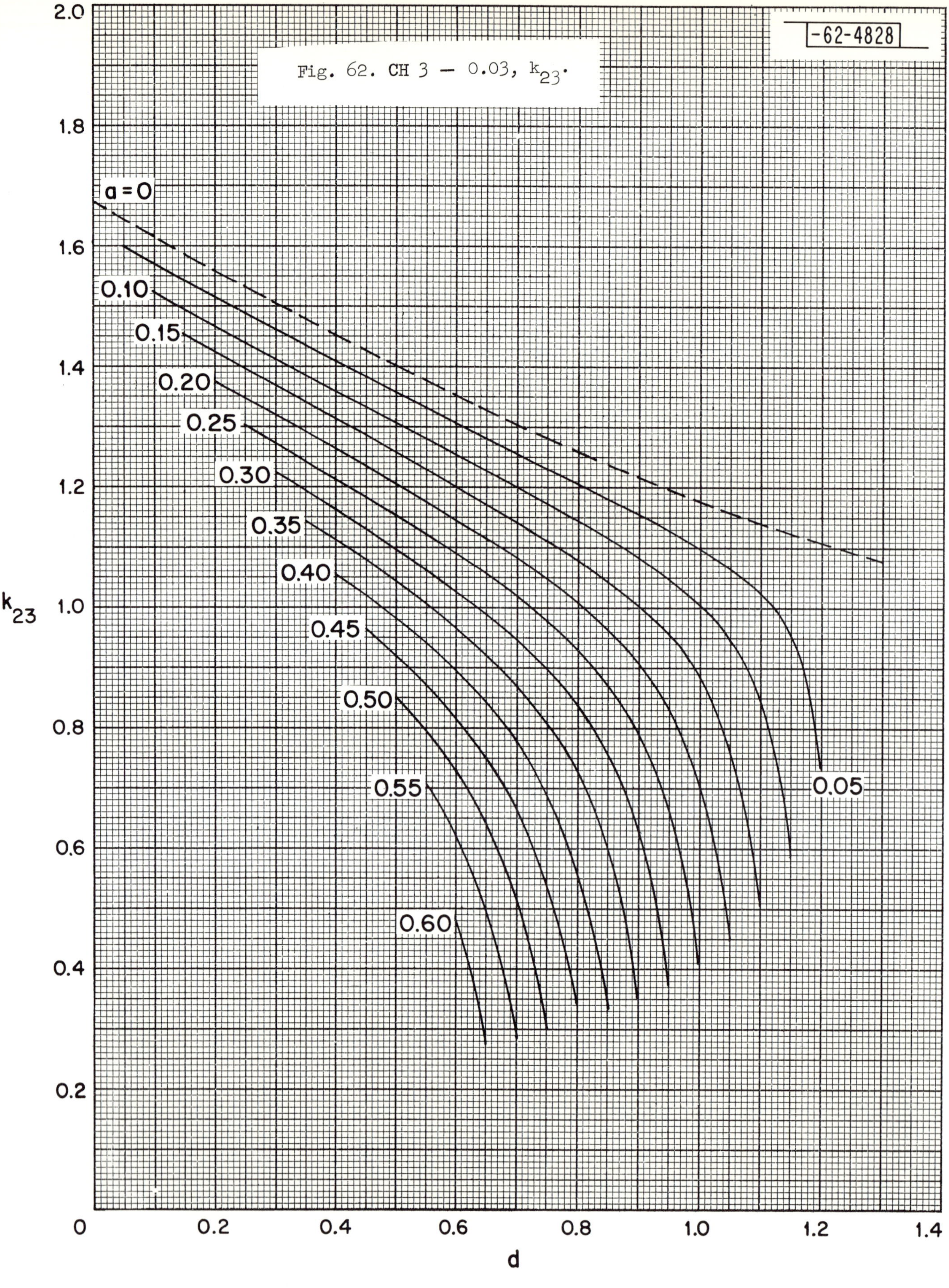


Fig. 62. CH 3 — 0.03, $k_{23}$.

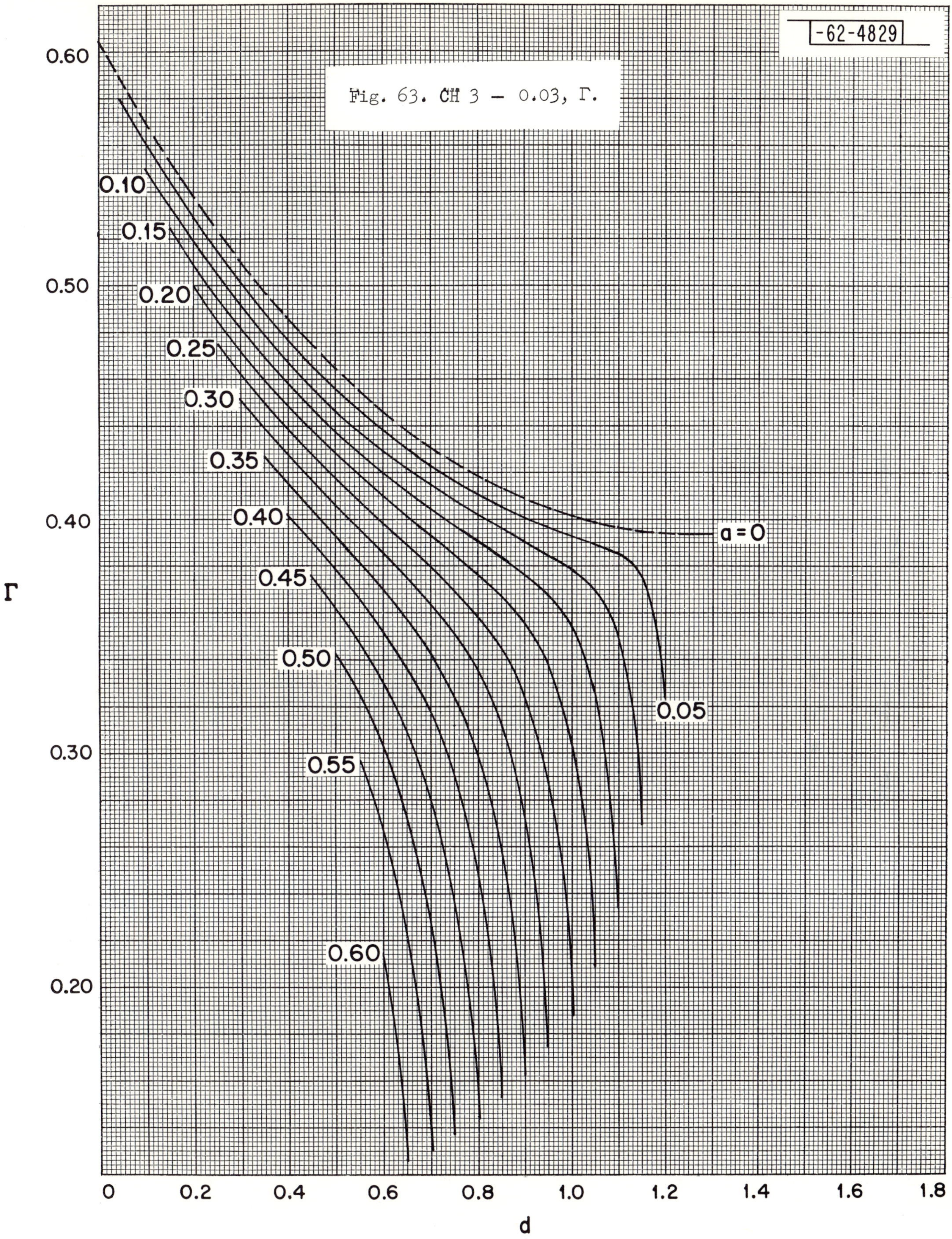

-62-4829
Fig. 63. CH 3 — 0.03, Γ.
0.60
0.50
0.40
0.30
0.20
Γ
0.10
0.15
0.20
0.25
0.30
0.35
0.40
0.45
0.50
0.55
0.60
a = 0
0.05
0
0.2
0.4
0.6
0.8
1.0
1.2
1.4
1.6
1.8
d

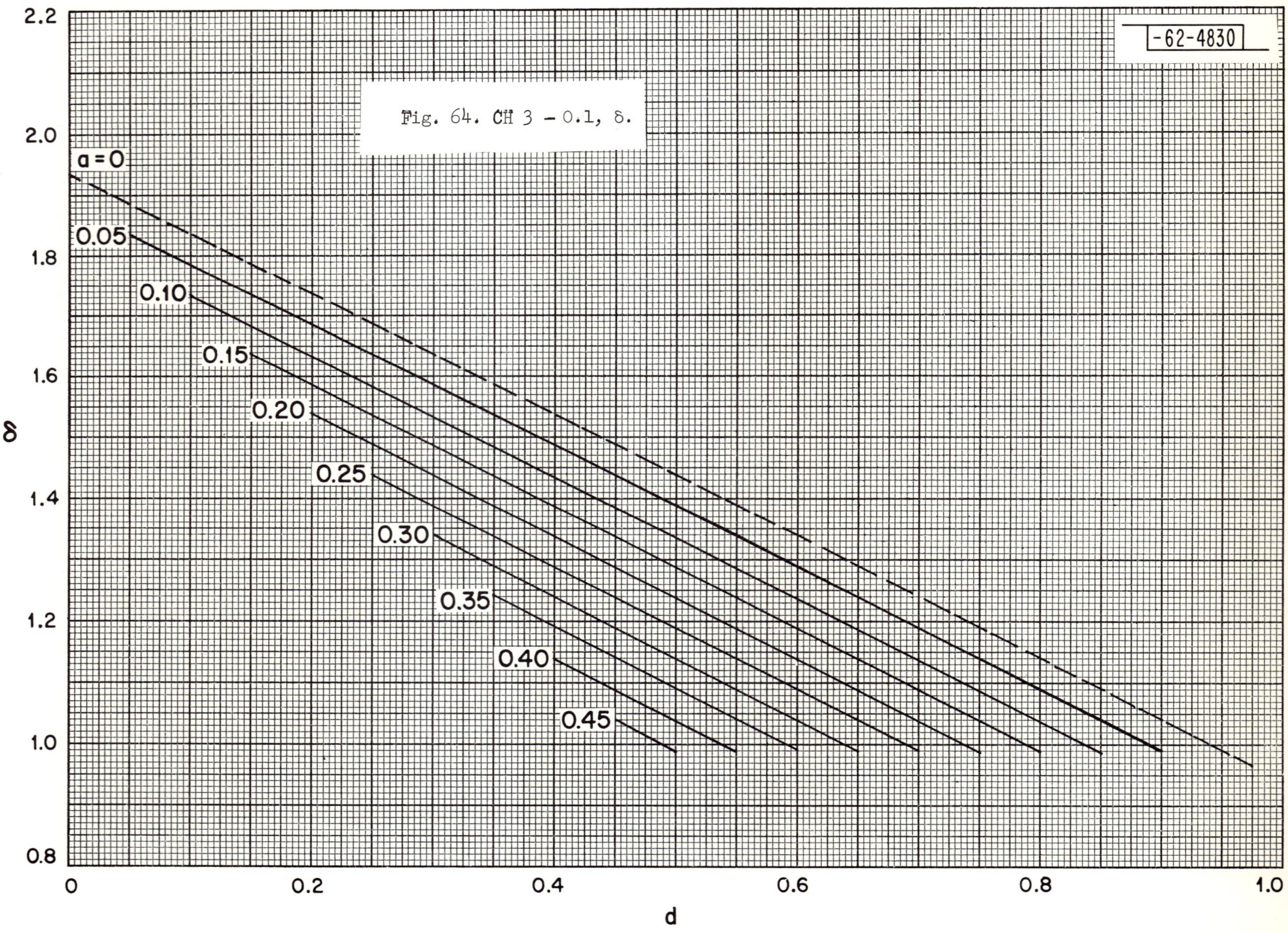


Fig. 64. CH 3 – 0.1, δ.

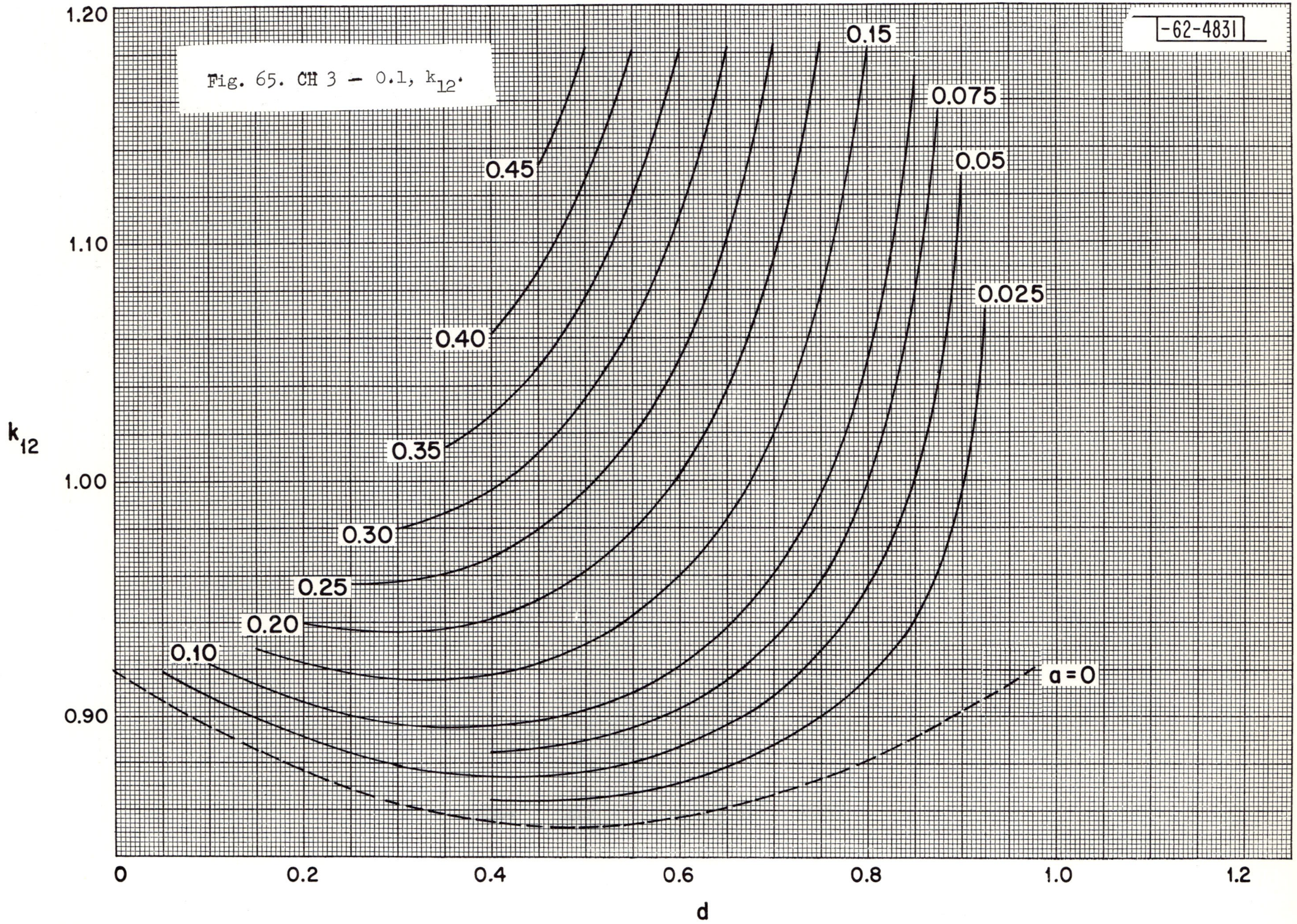


Fig. 65. CH 3 — 0.1, $k_{12}$.

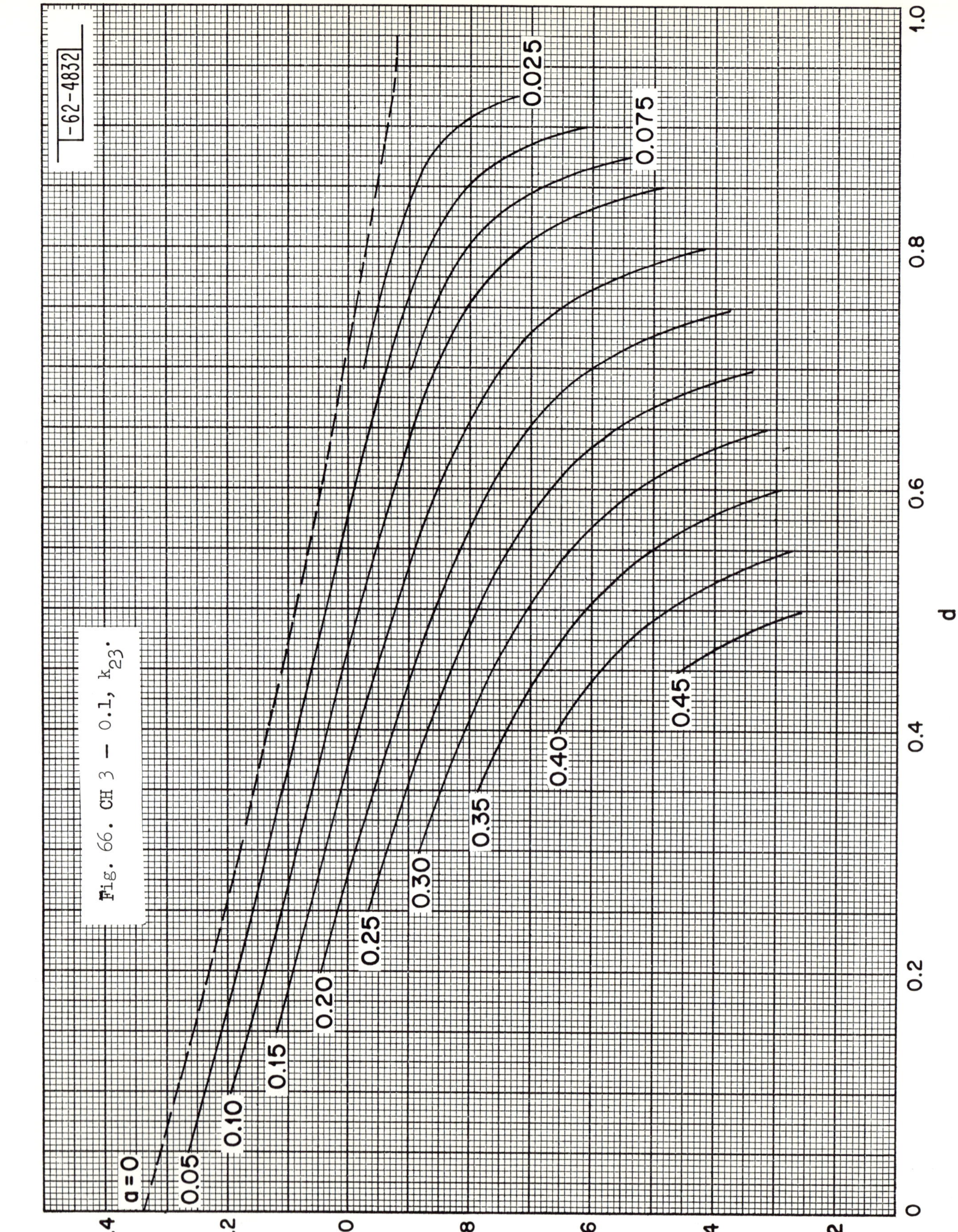


Fig. 66. CH 3 — 0.1, $k_{23}$.

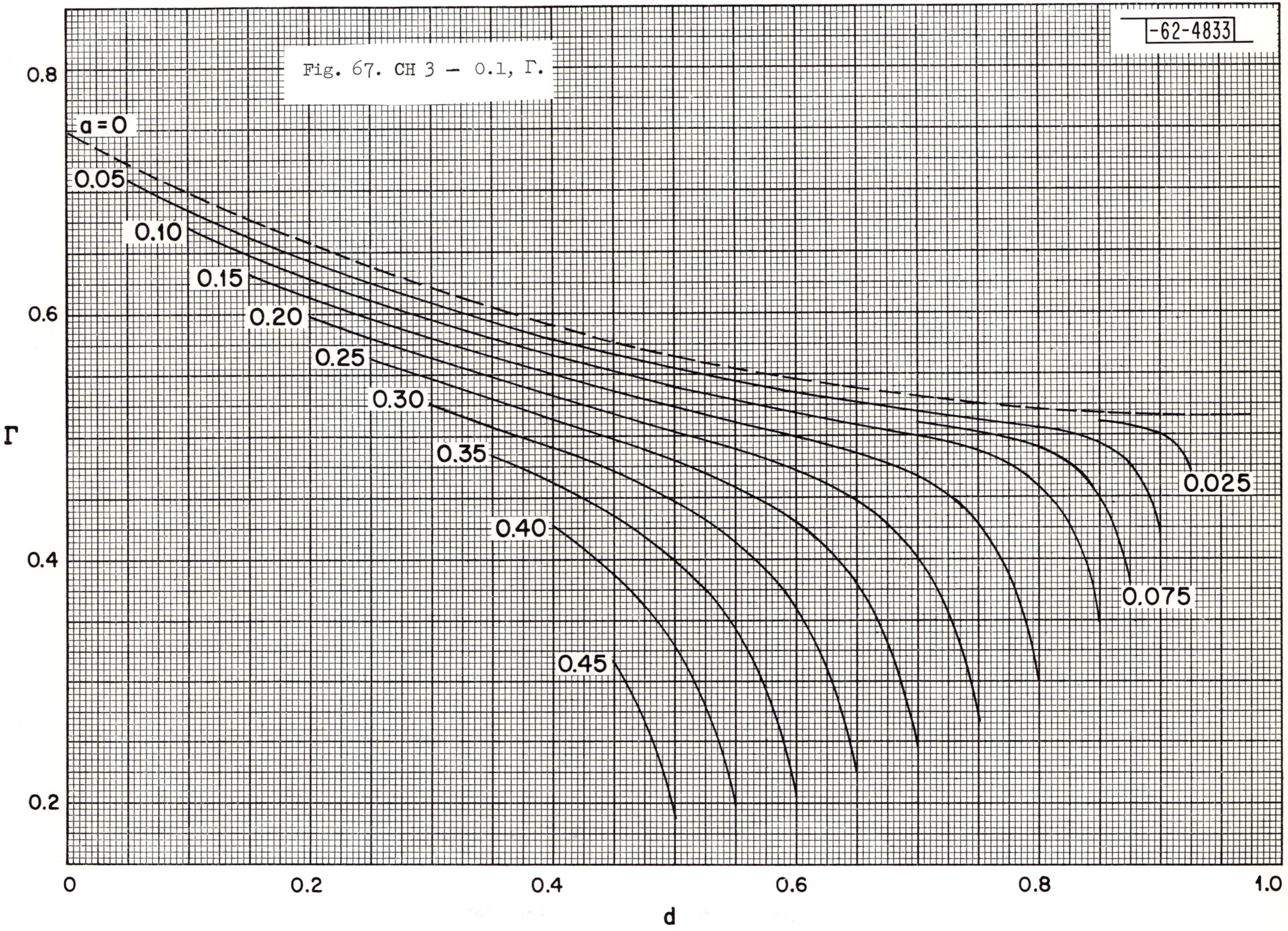


Fig. 67. CH 3 — 0.1, Γ.

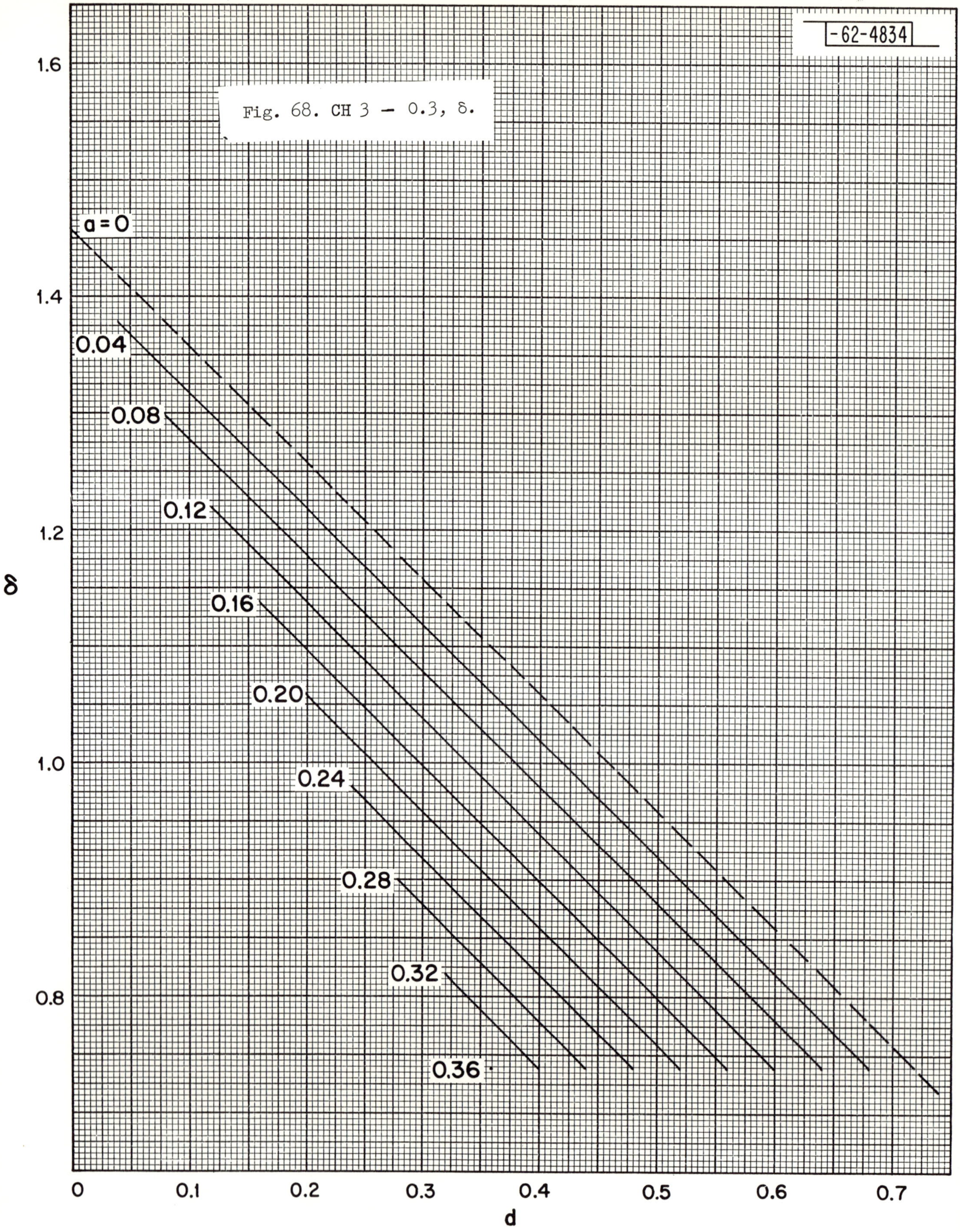


Fig. 68. CH 3 — 0.3, δ.

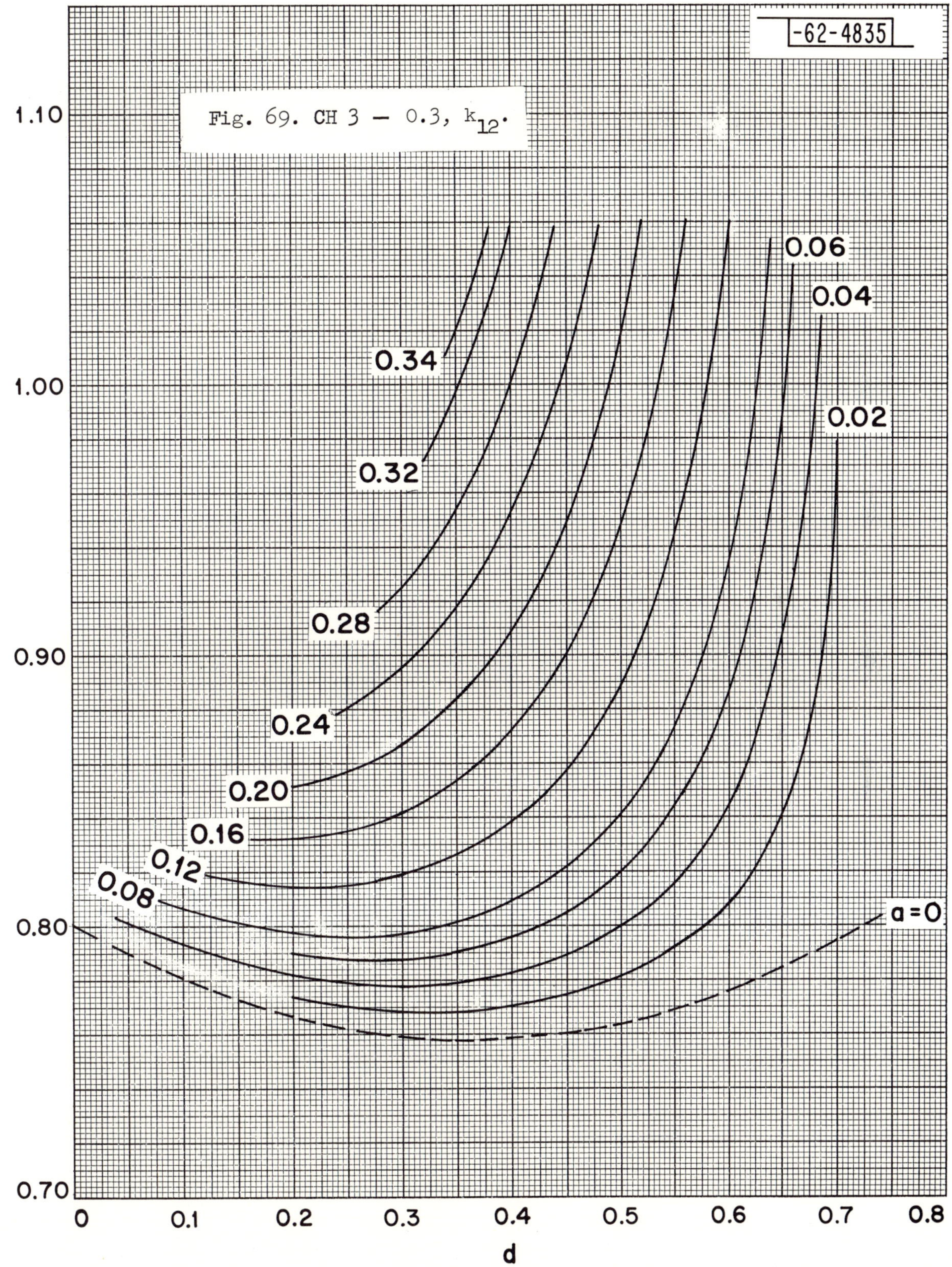

-62-4835
Fig. 69. CH 3 — 0.3, k12.
1.10
1.00
0.90
0.80
0.70
k12
0.34
0.32
0.28
0.24
0.20
0.16
0.12
0.08
0.06
0.04
0.02
a=0
0
0.1
0.2
0.3
0.4
0.5
0.6
0.7
0.8
d

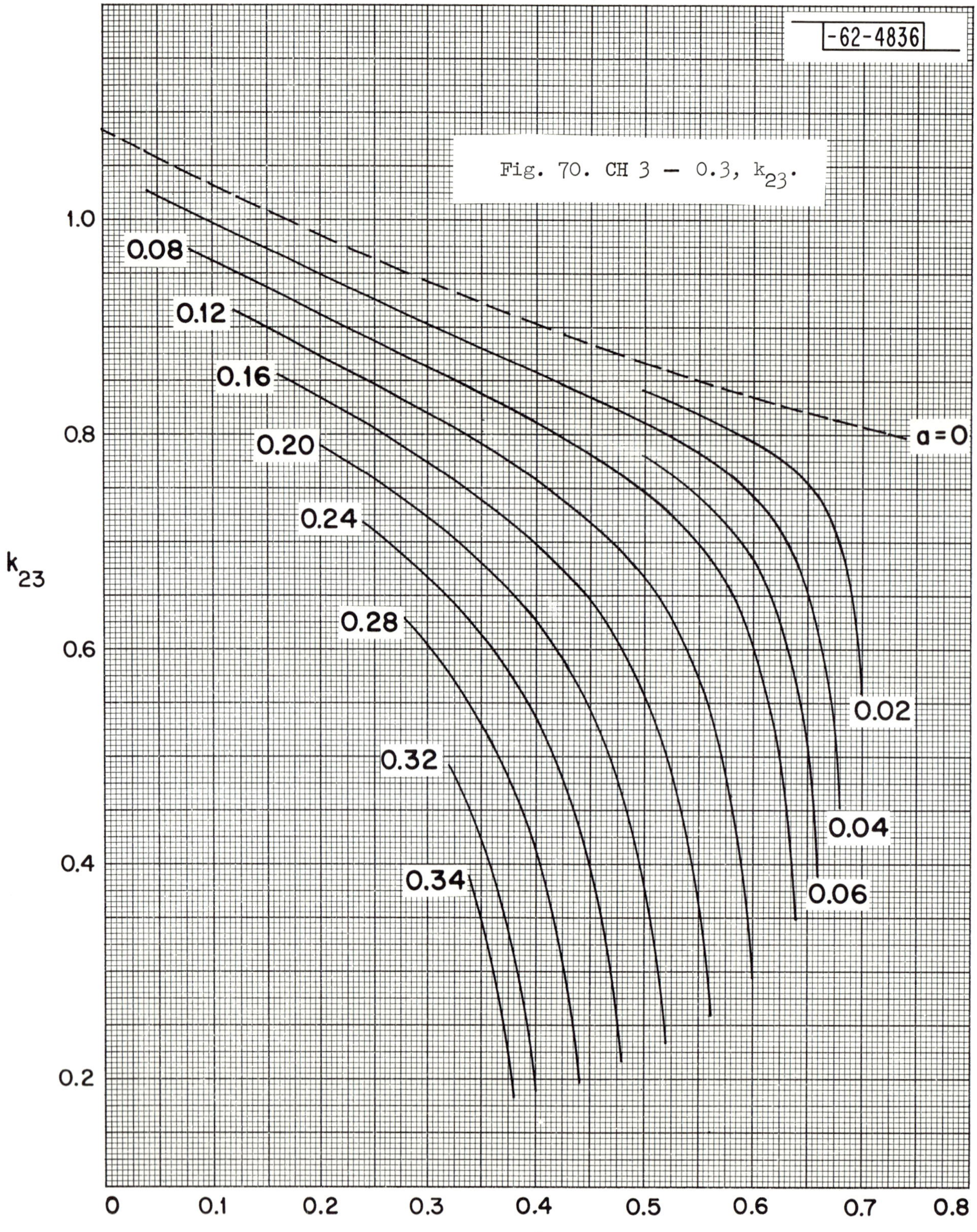
-62-4836
Fig. 70. CH 3 — 0.3, $k_{23}$.
1.0
0.8
0.6
0.4
0.2
$k_{23}$
0.08
0.12
0.16
0.20
0.24
0.28
0.32
0.34
0.02
0.04
0.06
a=0
0
0.1
0.2
0.3
0.4
0.5
0.6
0.7
0.8
d

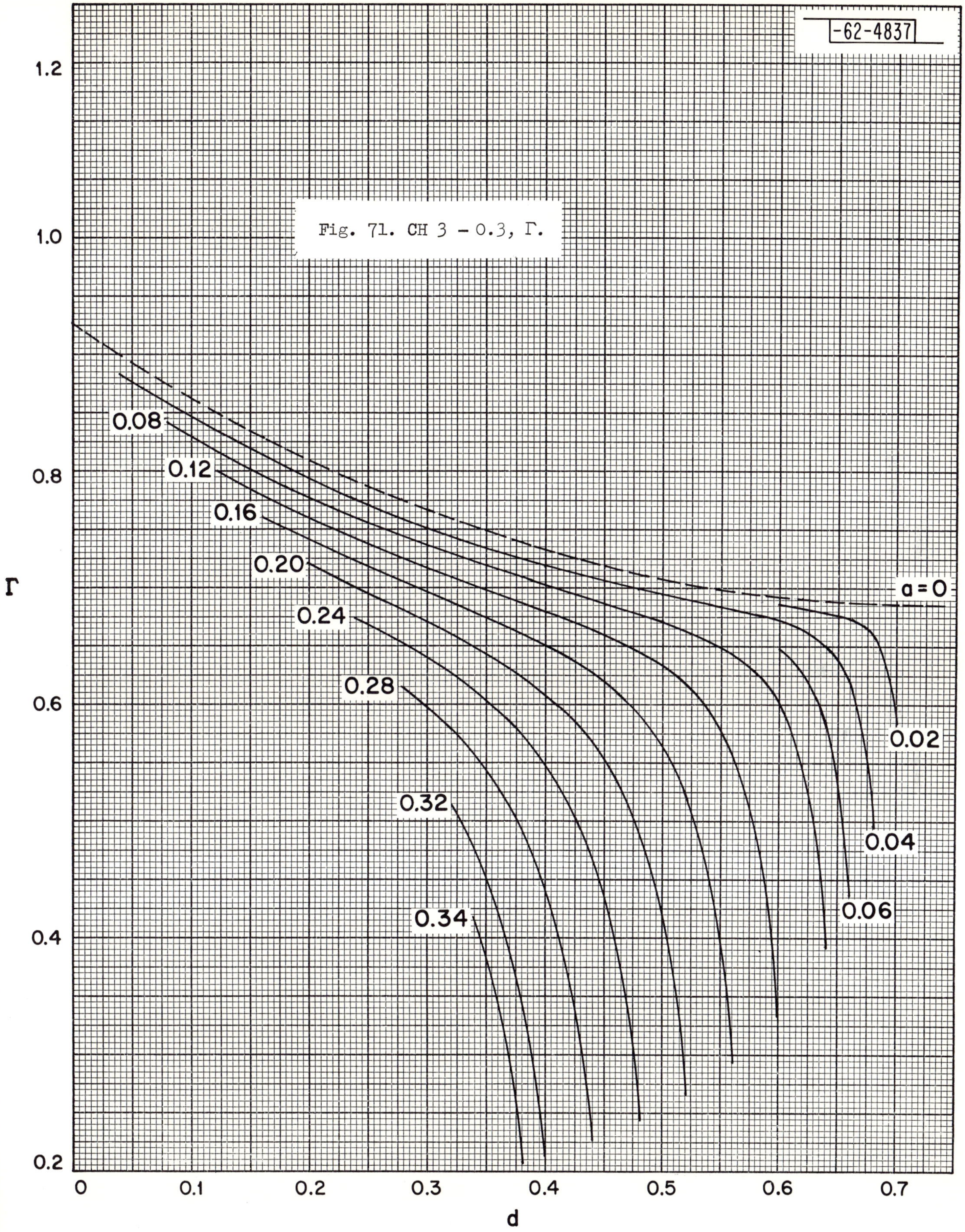
-62-4837
Fig. 71. CH 3 – 0.3, Γ.
1.2
1.0
0.8
0.6
0.4
0.2
Γ
0
0.1
0.2
0.3
0.4
0.5
0.6
0.7
d
0.08
0.12
0.16
0.20
0.24
0.28
0.32
0.34
a = 0
0.02
0.04
0.06

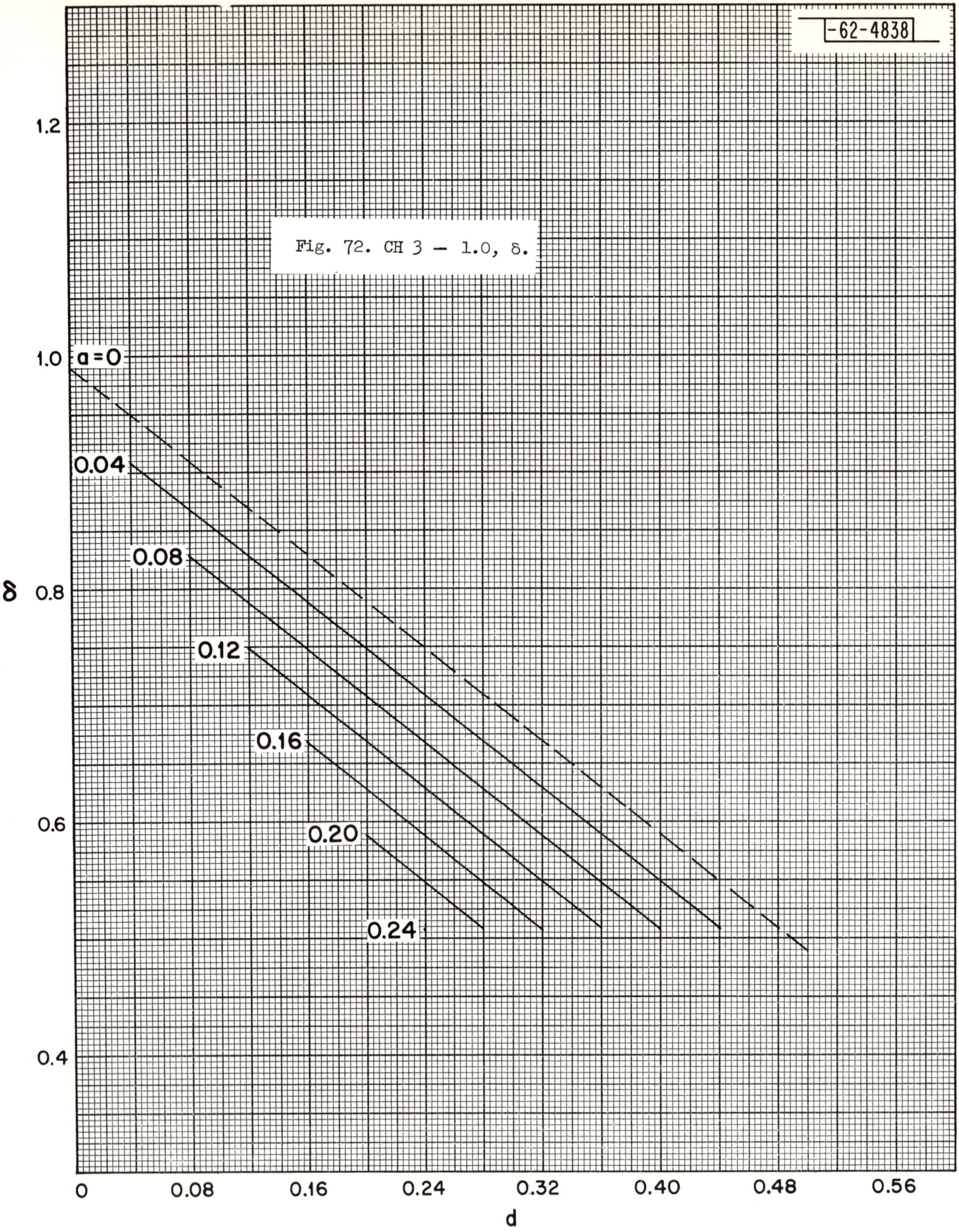


Fig. 72. CH 3 — 1.0, δ.

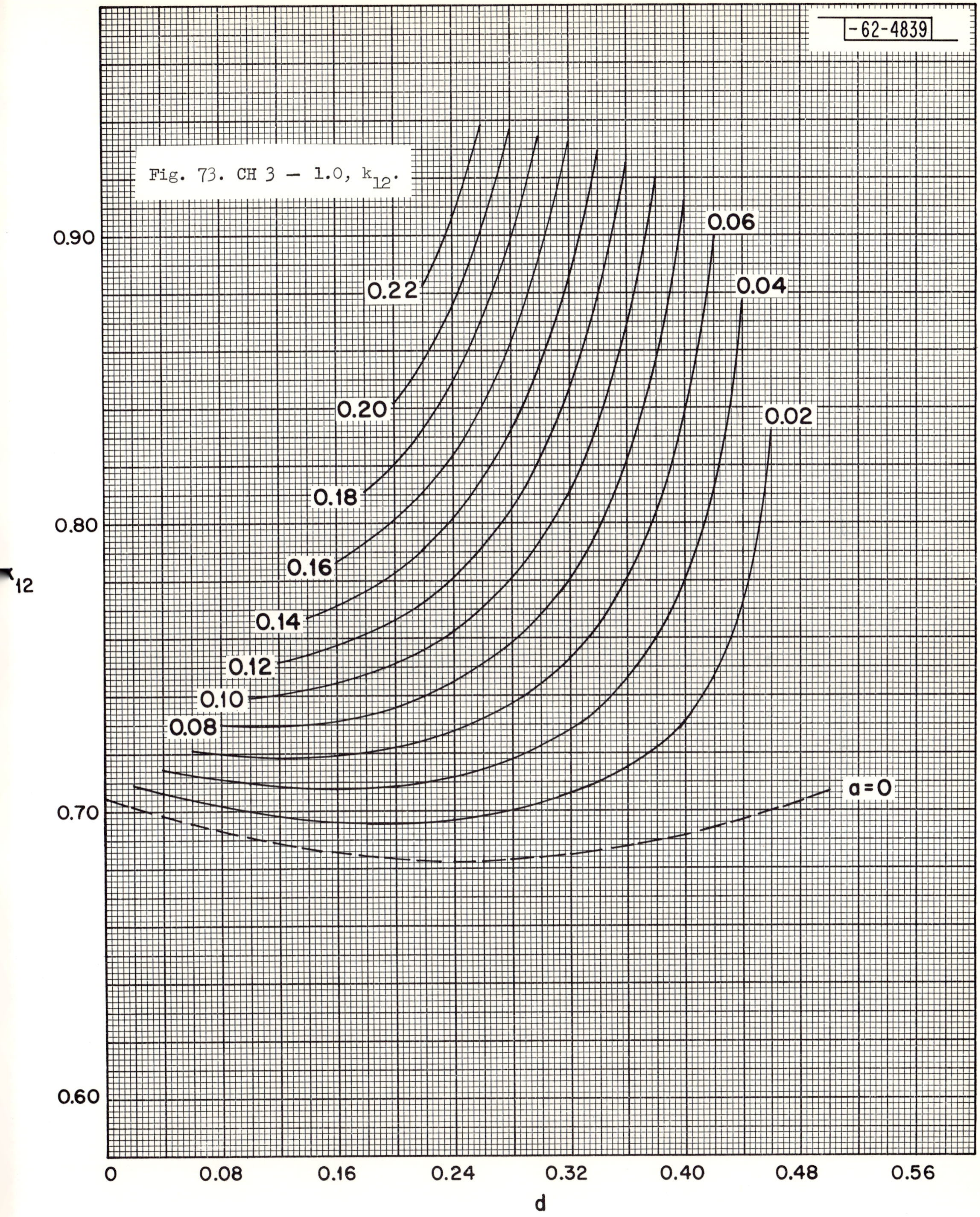
-62-4839
Fig. 73. CH 3 — 1.0, k12.
0.90
0.80
0.70
0.60
k12
0
0.08
0.16
0.24
0.32
0.40
0.48
0.56
d
0.22
0.20
0.18
0.16
0.14
0.12
0.10
0.08
0.06
0.04
0.02
a=0

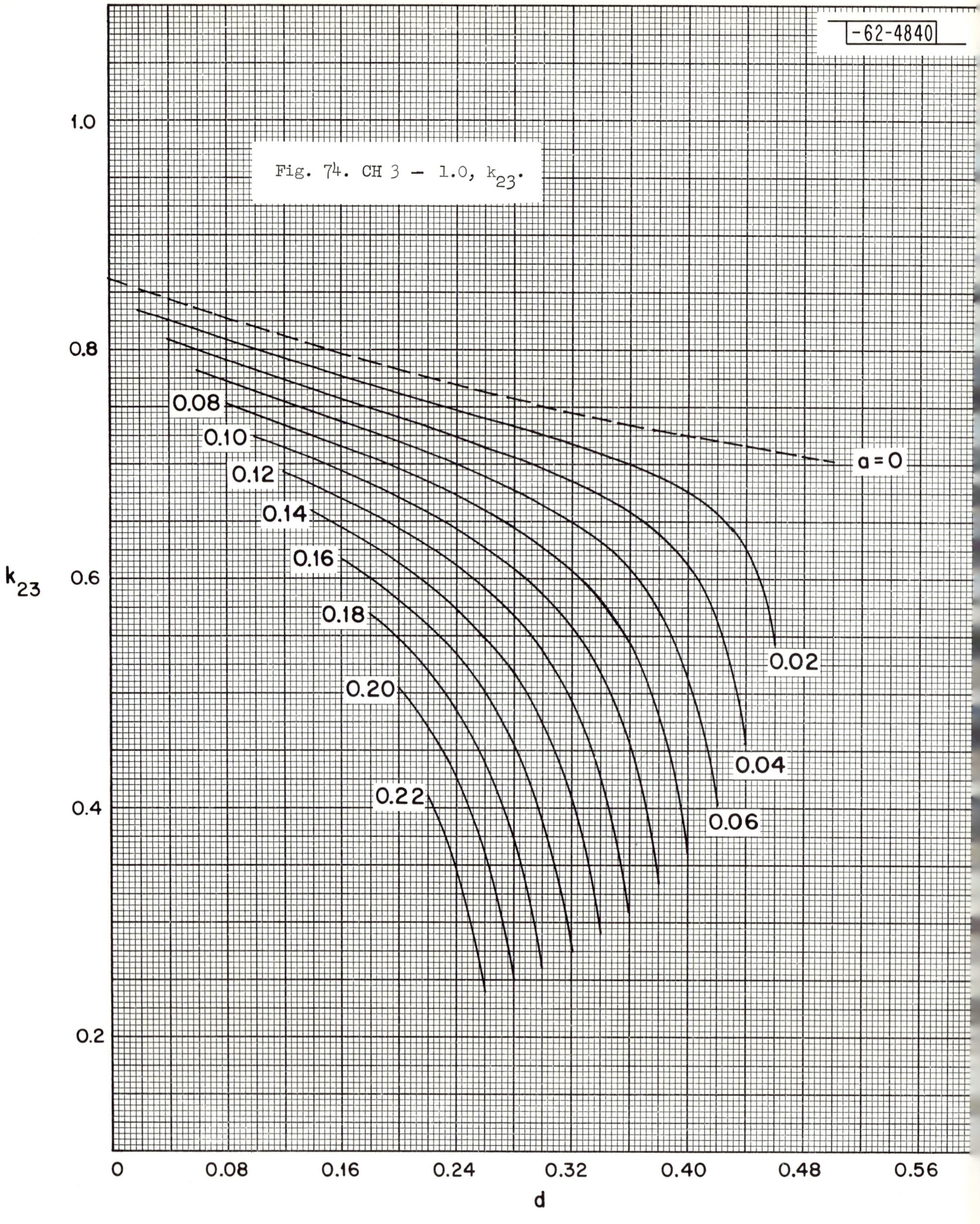
-62-4840
Fig. 74. CH 3 — 1.0, $k_{23}$.
1.0
0.8
0.6
0.4
0.2
$k_{23}$
0
0.08
0.16
0.24
0.32
0.40
0.48
0.56
d
a = 0
0.02
0.04
0.06
0.08
0.10
0.12
0.14
0.16
0.18
0.20
0.22

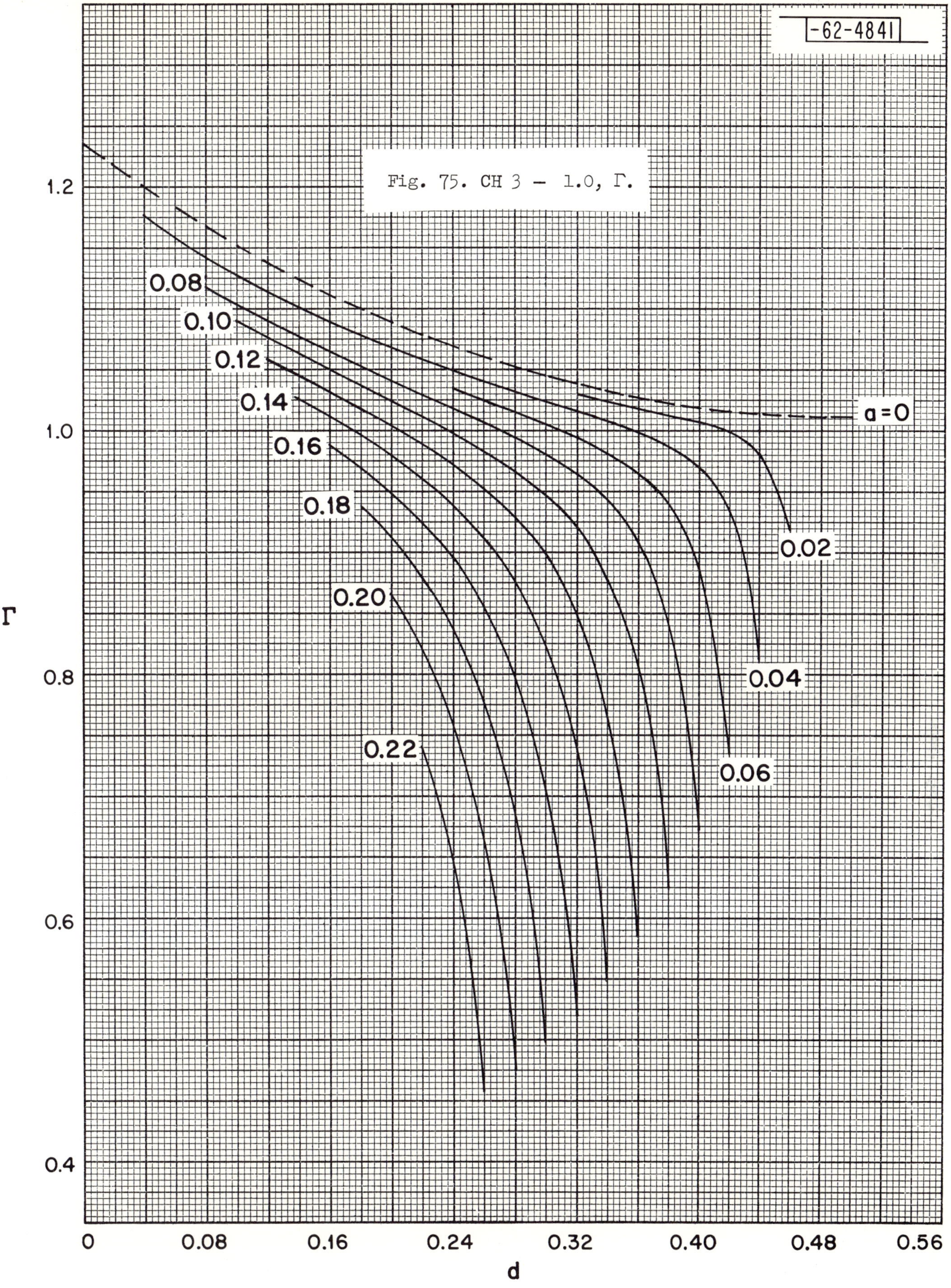

-62-4841
Fig. 75. CH 3 — 1.0, Γ.
1.2
1.0
0.8
0.6
0.4
Γ
0
0.08
0.16
0.24
0.32
0.40
0.48
0.56
d
a=0
0.02
0.04
0.06
0.08
0.10
0.12
0.14
0.16
0.18
0.20
0.22

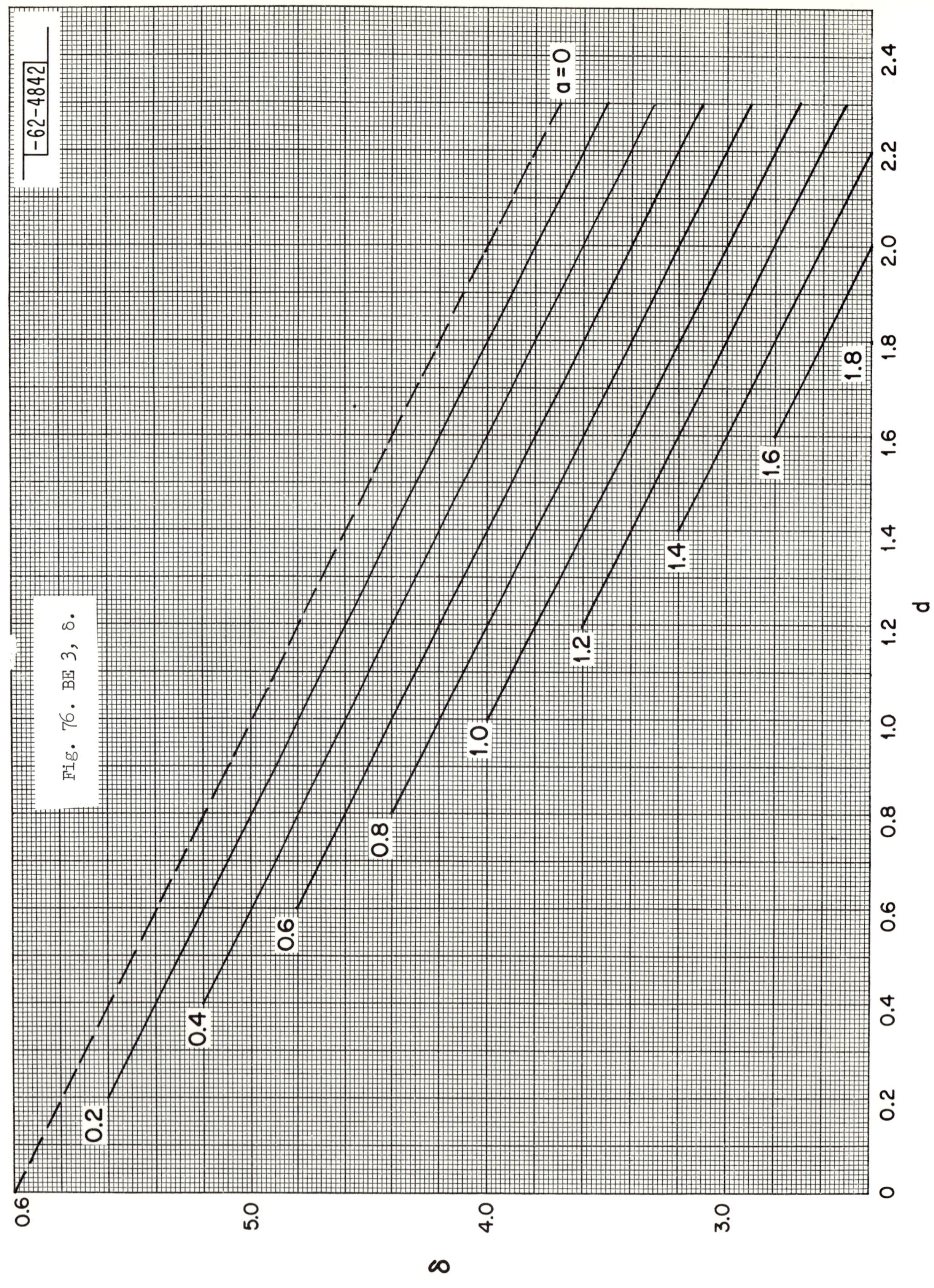
-62-4842
a=0
0.2
0.4
0.6
0.8
1.0
1.2
1.4
1.6
1.8
δ
6.0
5.0
4.0
3.0
d
0
0.2
0.4
0.6
0.8
1.0
1.2
1.4
1.6
1.8
2.0
2.2
2.4

Fig. 76. BE 3, δ.

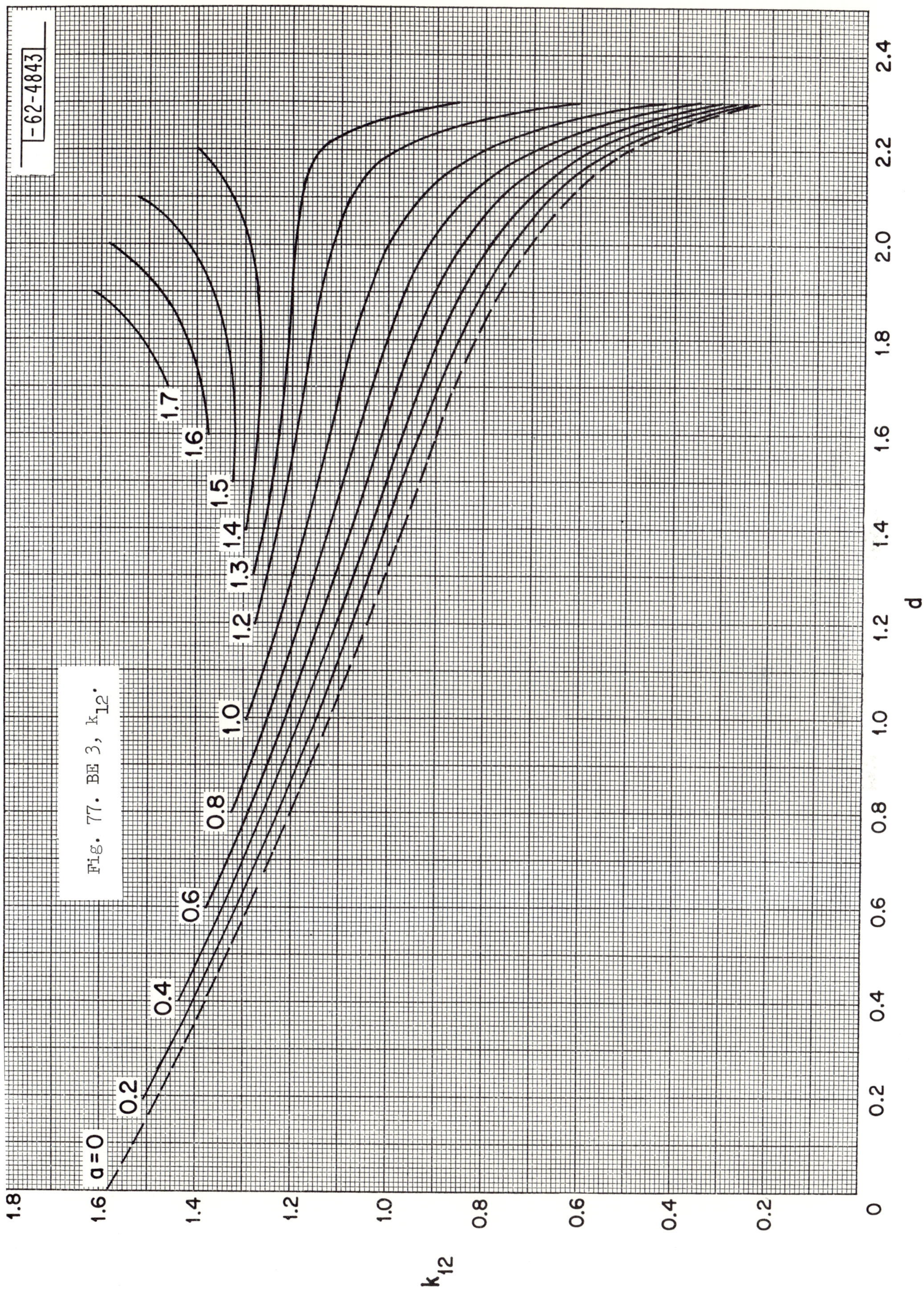


Fig. 77. BE 3, $k_{12}$.

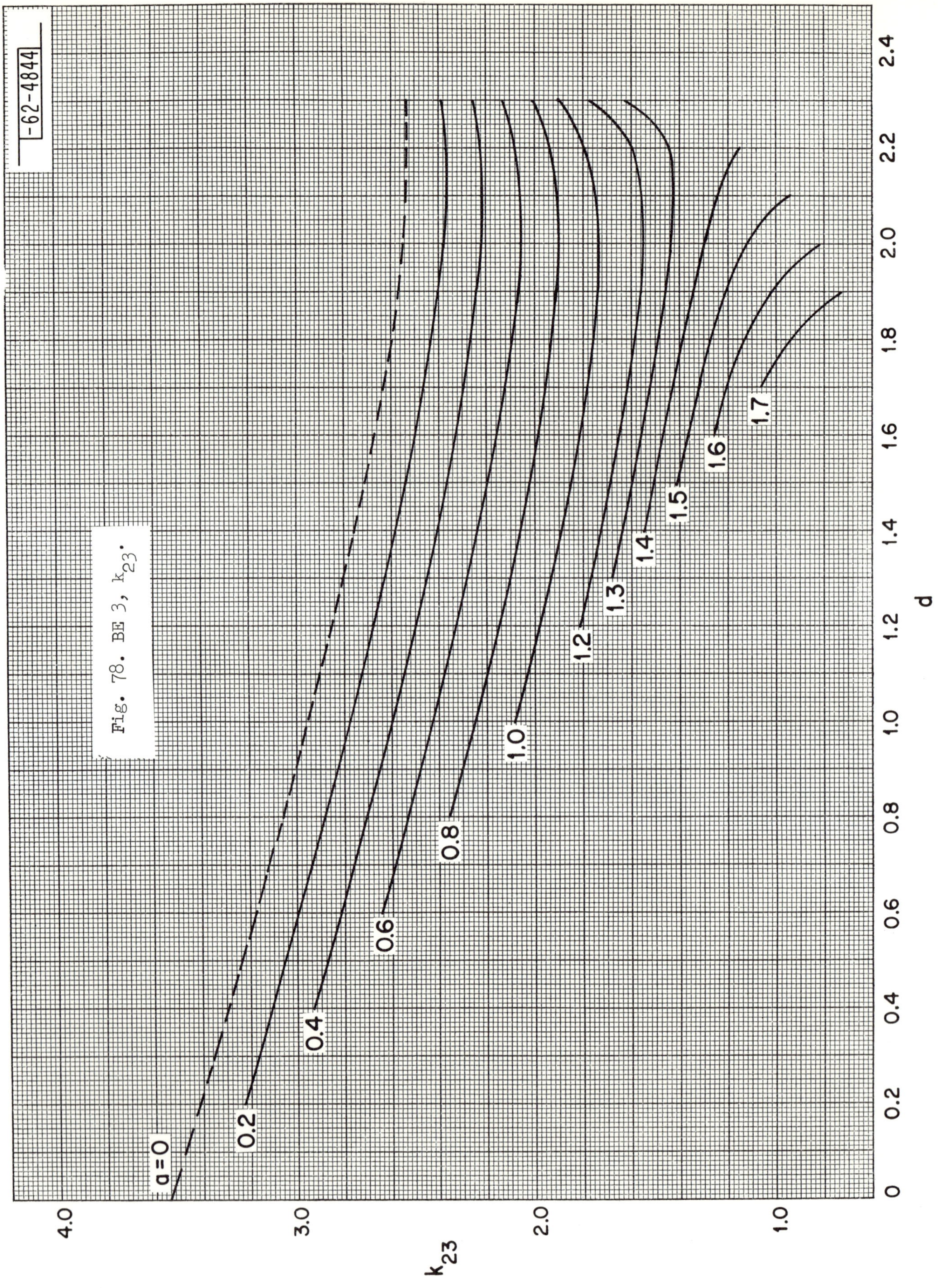


Fig. 78. BE 3, $k_{23}$.

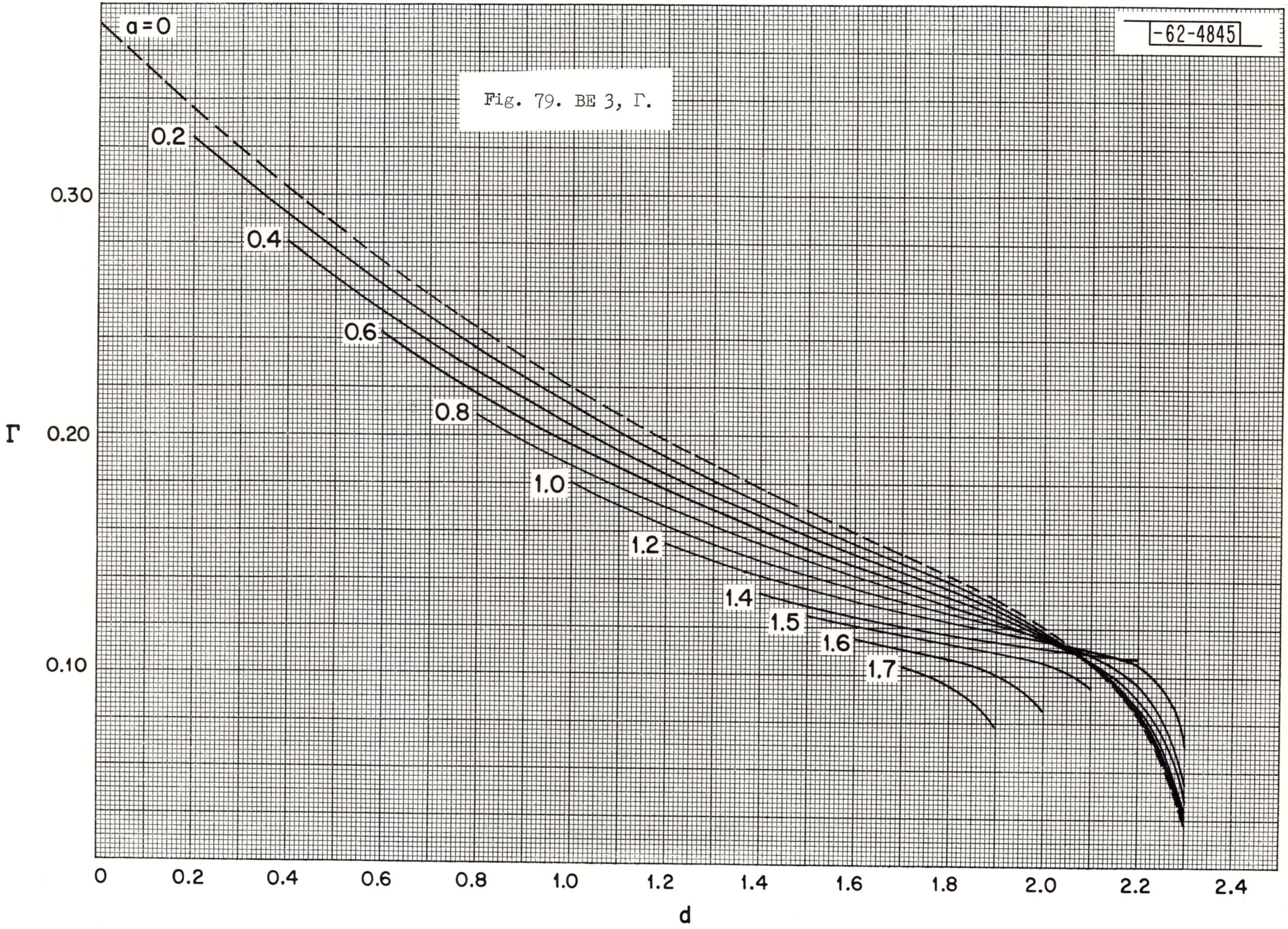
a=0
-62-4845
0.2
0.4
0.6
0.8
1.0
1.2
1.4
1.5
1.6
1.7
0.30
0.20
0.10
Γ
0
0.2
0.4
0.6
0.8
1.0
1.2
1.4
1.6
1.8
2.0
2.2
2.4
d

Fig. 79. BE 3, Γ.

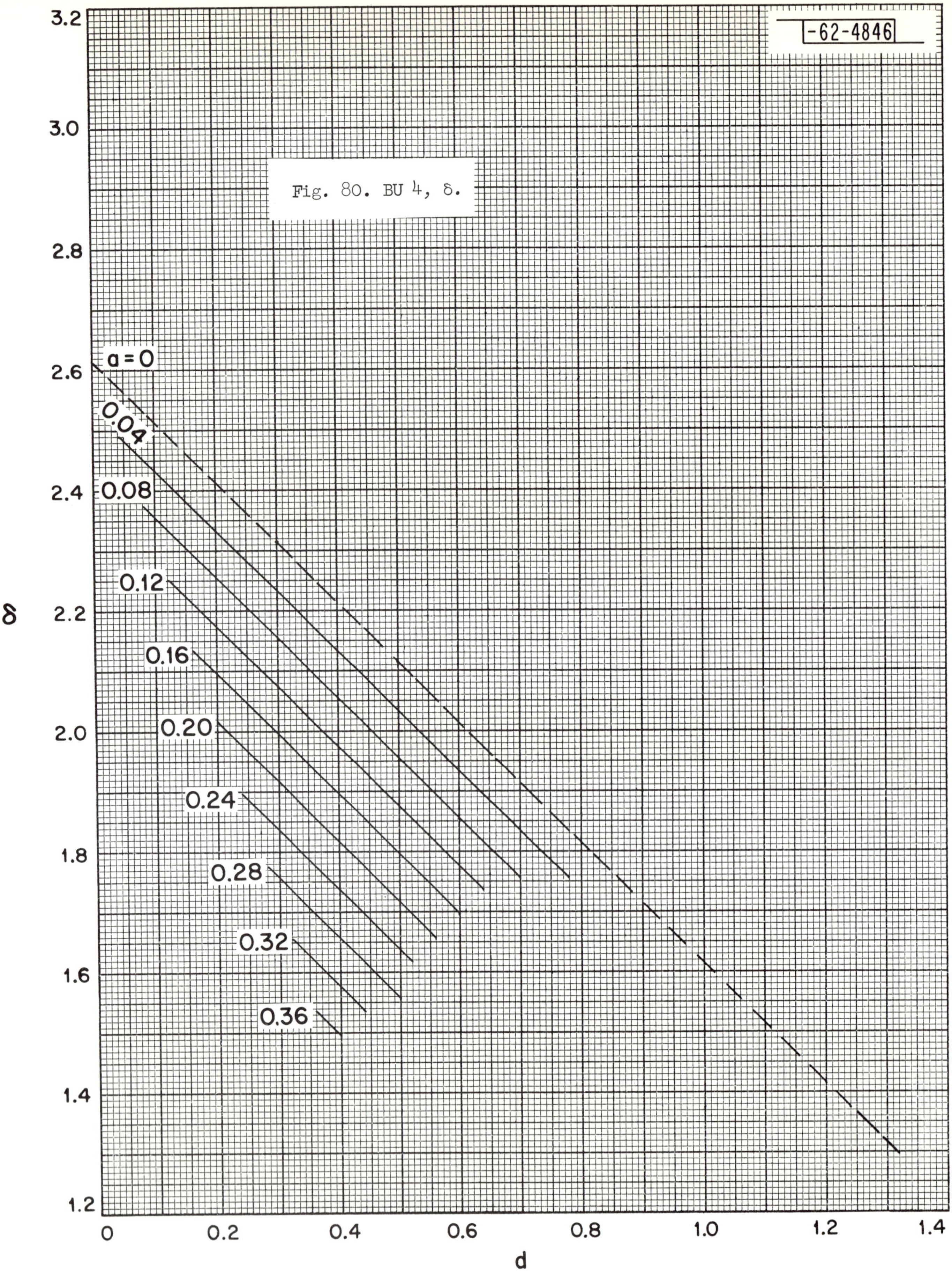
-62-4846
3.2
3.0
2.8
2.6
2.4
2.2
2.0
1.8
1.6
1.4
1.2
δ
a = 0
0.04
0.08
0.12
0.16
0.20
0.24
0.28
0.32
0.36
0
0.2
0.4
0.6
0.8
1.0
1.2
1.4
d

Fig. 80. BU 4, δ.

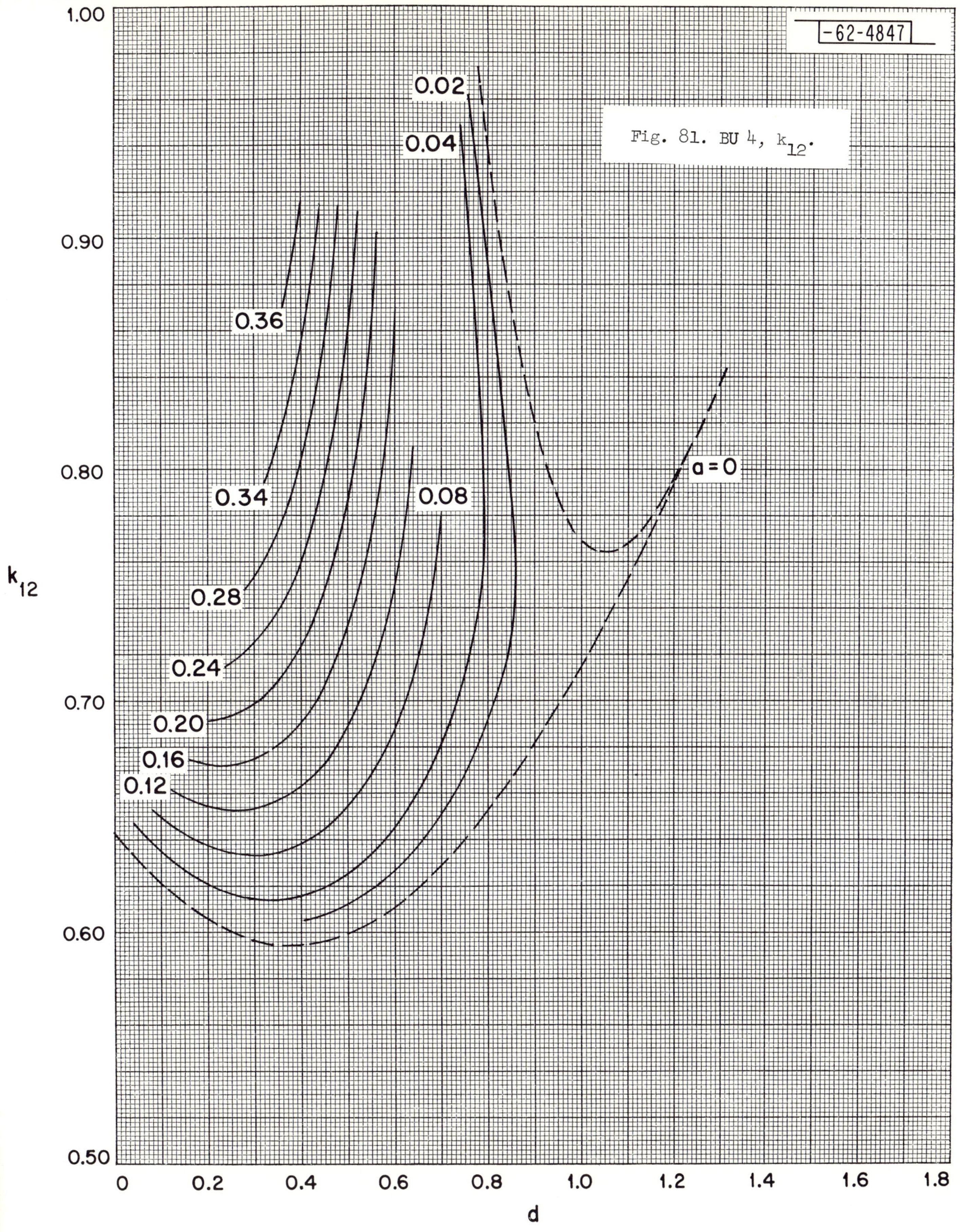
-62-4847
Fig. 81. BU 4, k12.
1.00
0.90
0.80
0.70
0.60
0.50
k12
0
0.2
0.4
0.6
0.8
1.0
1.2
1.4
1.6
1.8
d
0.02
0.04
0.36
0.34
0.08
0.28
0.24
0.20
0.16
0.12
a = 0

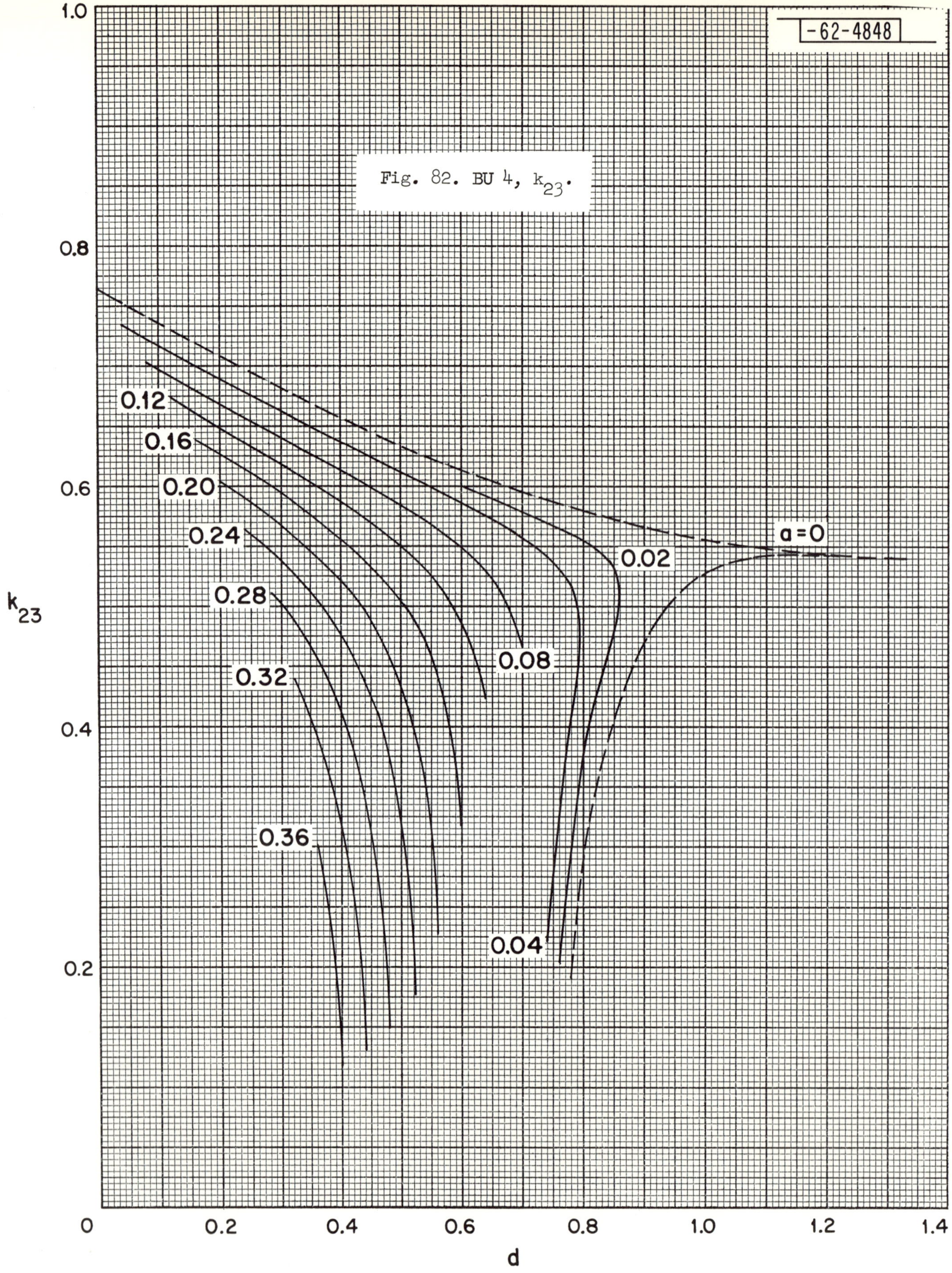


Fig. 82. BU 4, $k_{23}$.

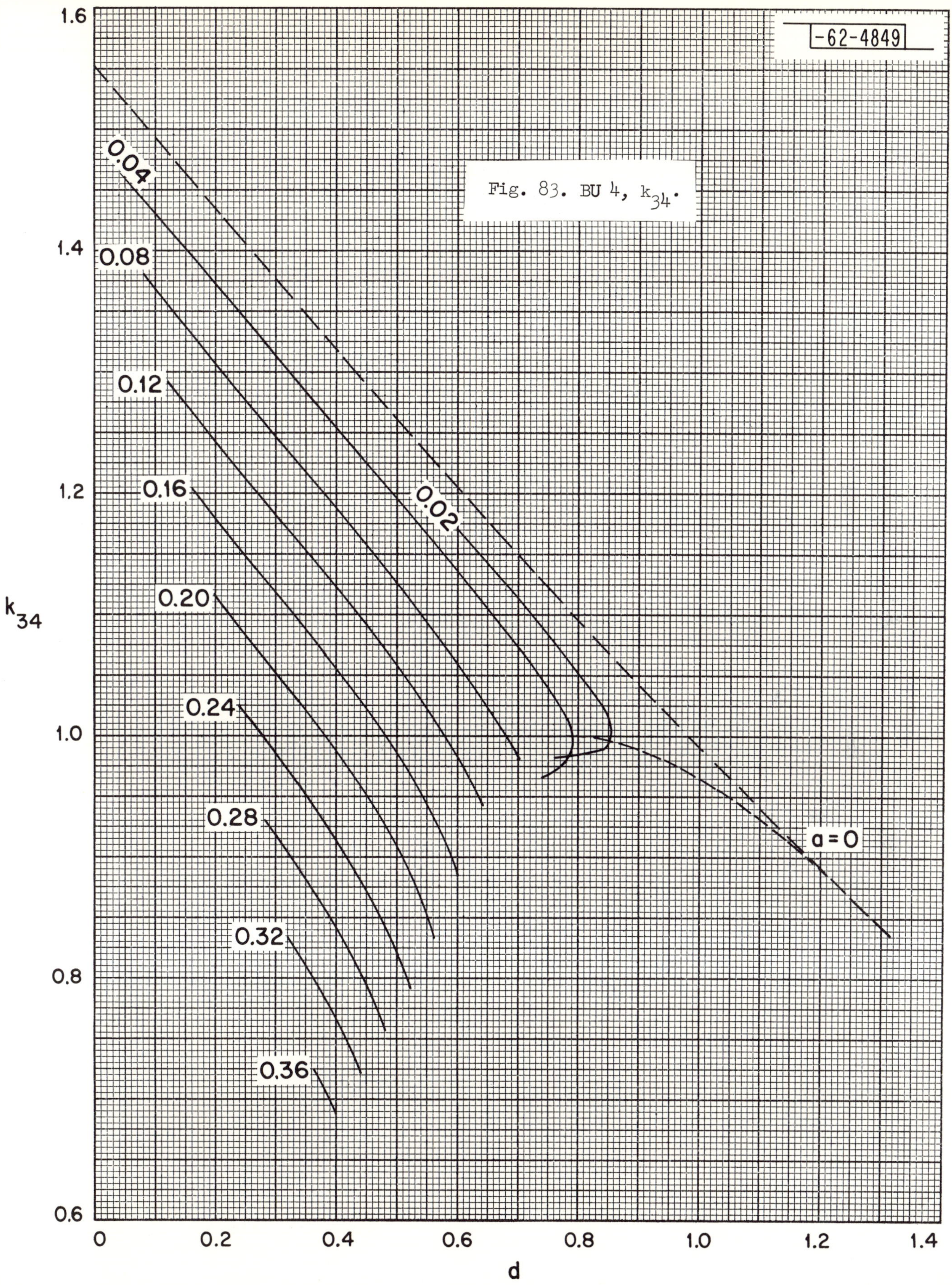


Fig. 83. BU 4, $k_{34}$.

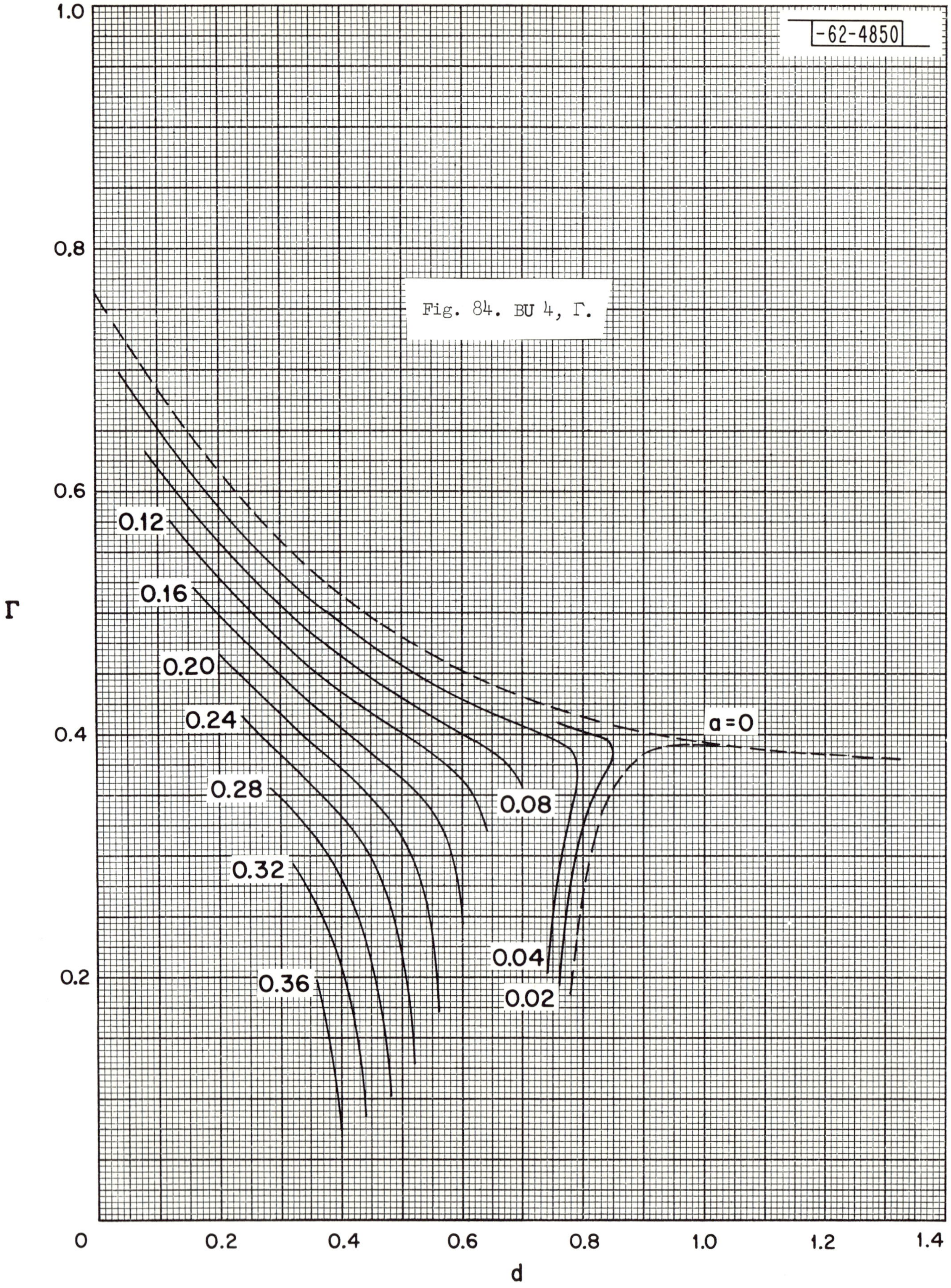

-62-4850
1.0
0.8
0.6
0.4
0.2
Γ
0.12
0.16
0.20
0.24
0.28
0.32
0.36
0.08
0.04
0.02
a=0
0
0.2
0.4
0.6
0.8
1.0
1.2
1.4
d

Fig. 84. BU 4, Γ.

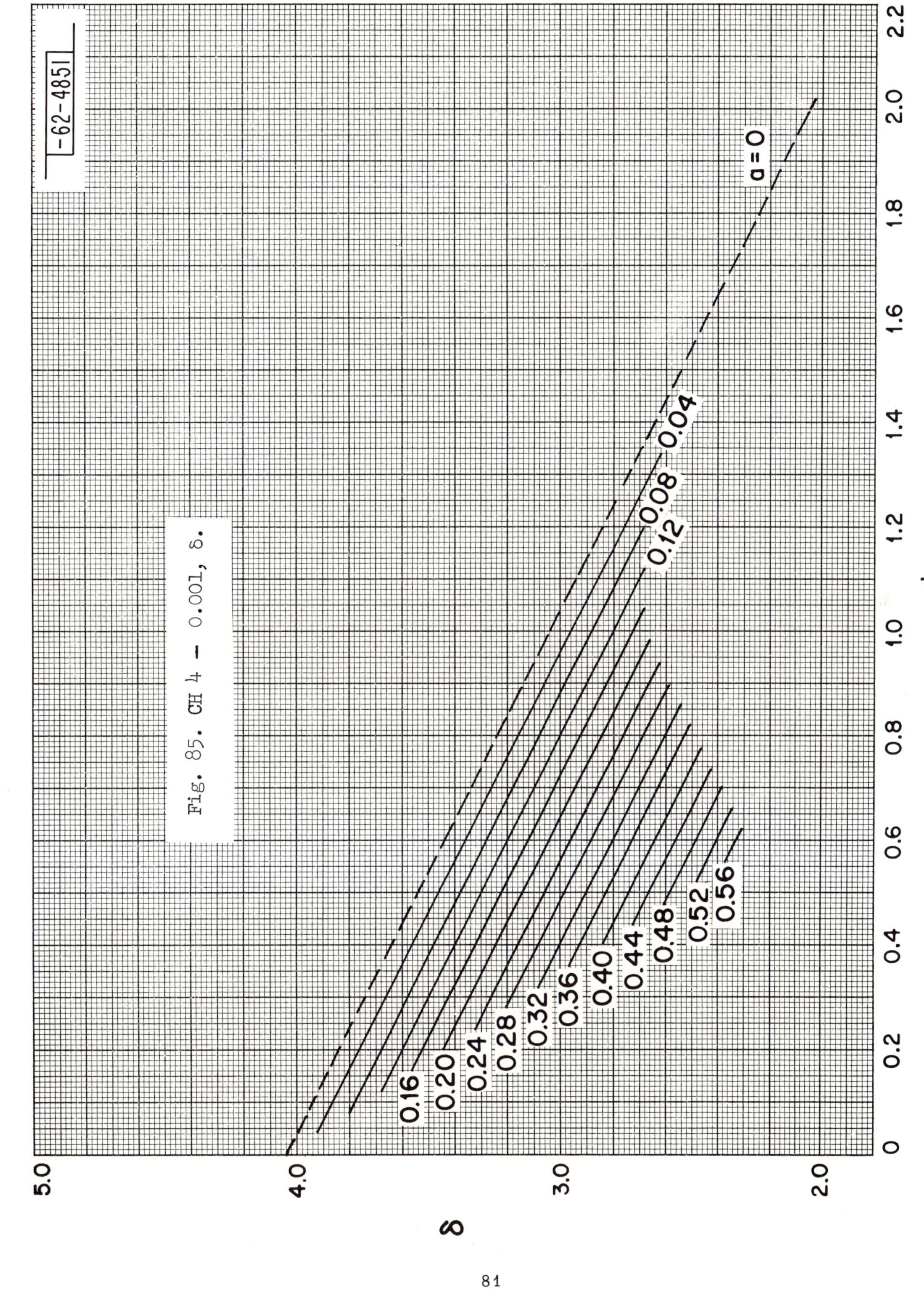


Fig. 85. CH 4 — 0.001, δ.

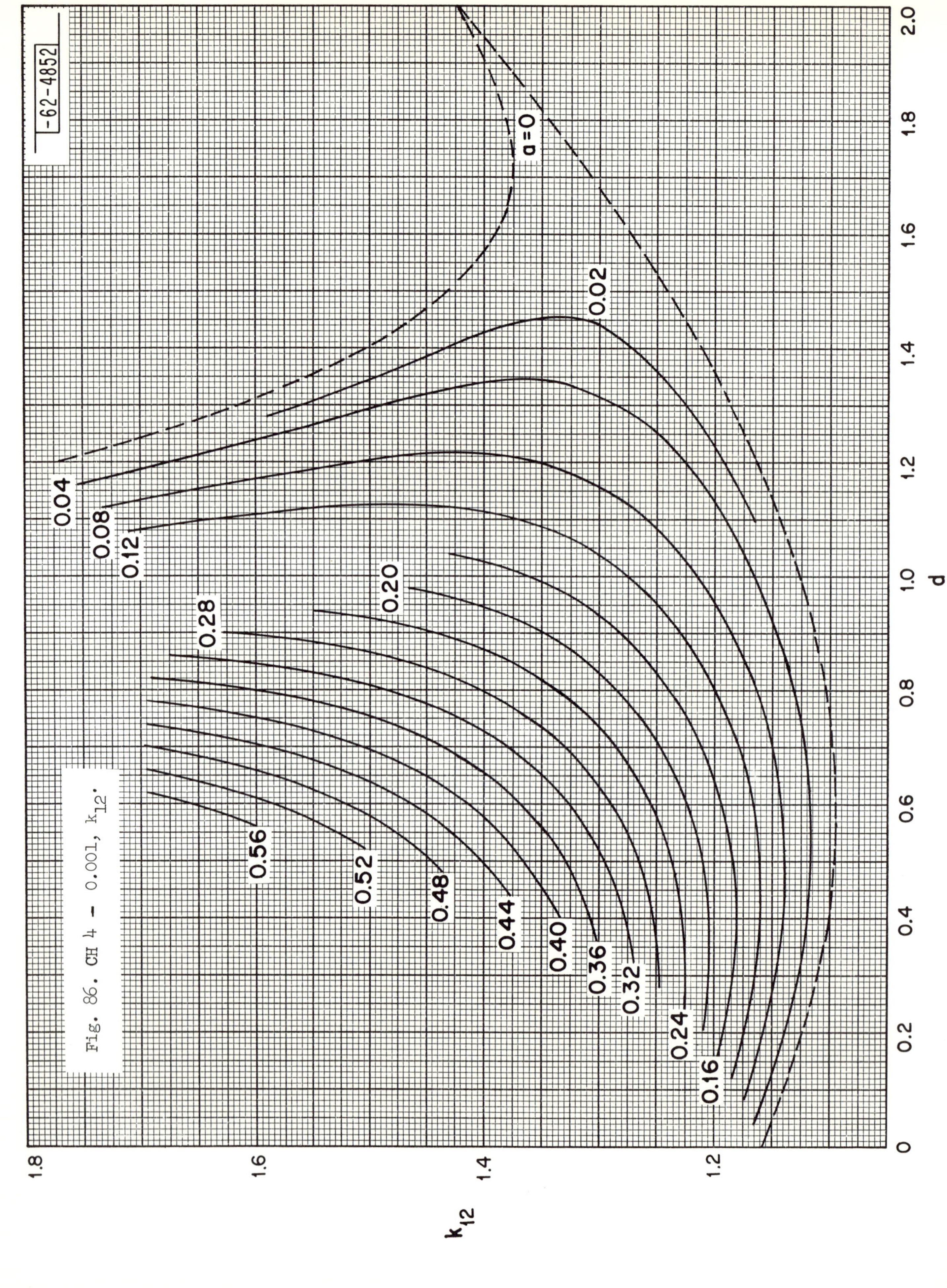


Fig. 86. CH 4 - 0.001, $k_{12}$.

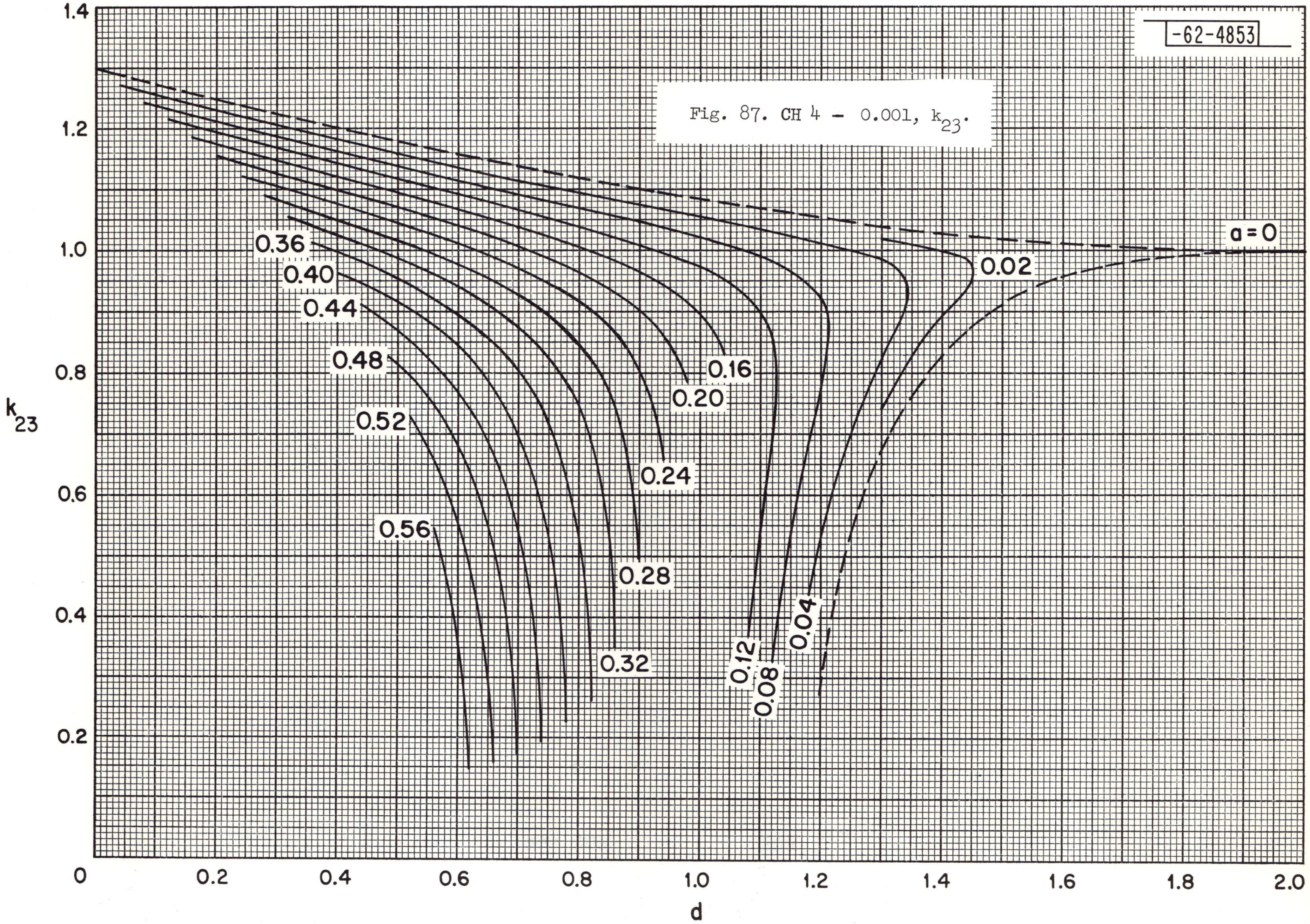


Fig. 87. CH 4 - 0.001, $k_{23}$.

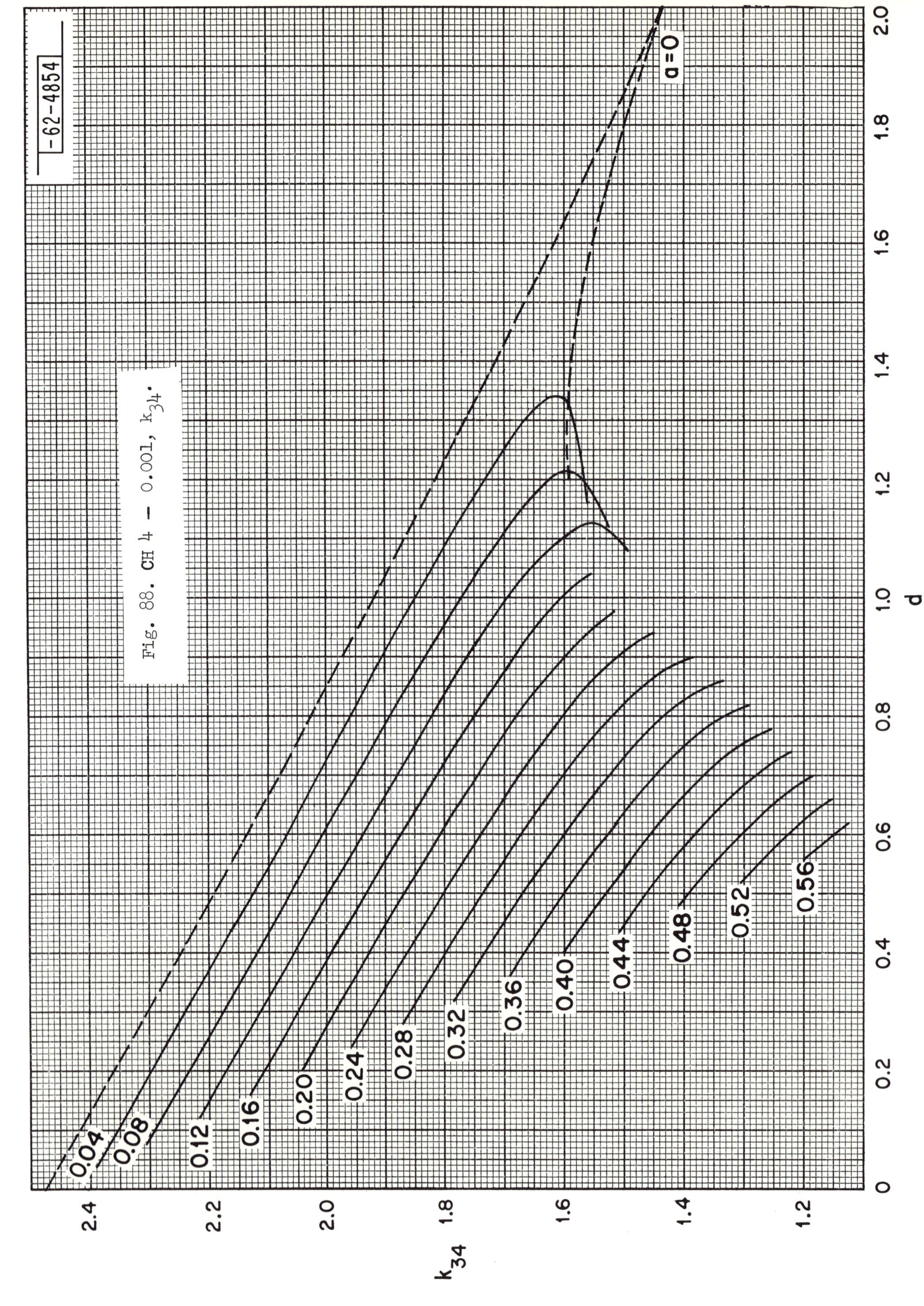


Fig. 88. CH 4 — 0.001, $k_{34}$.

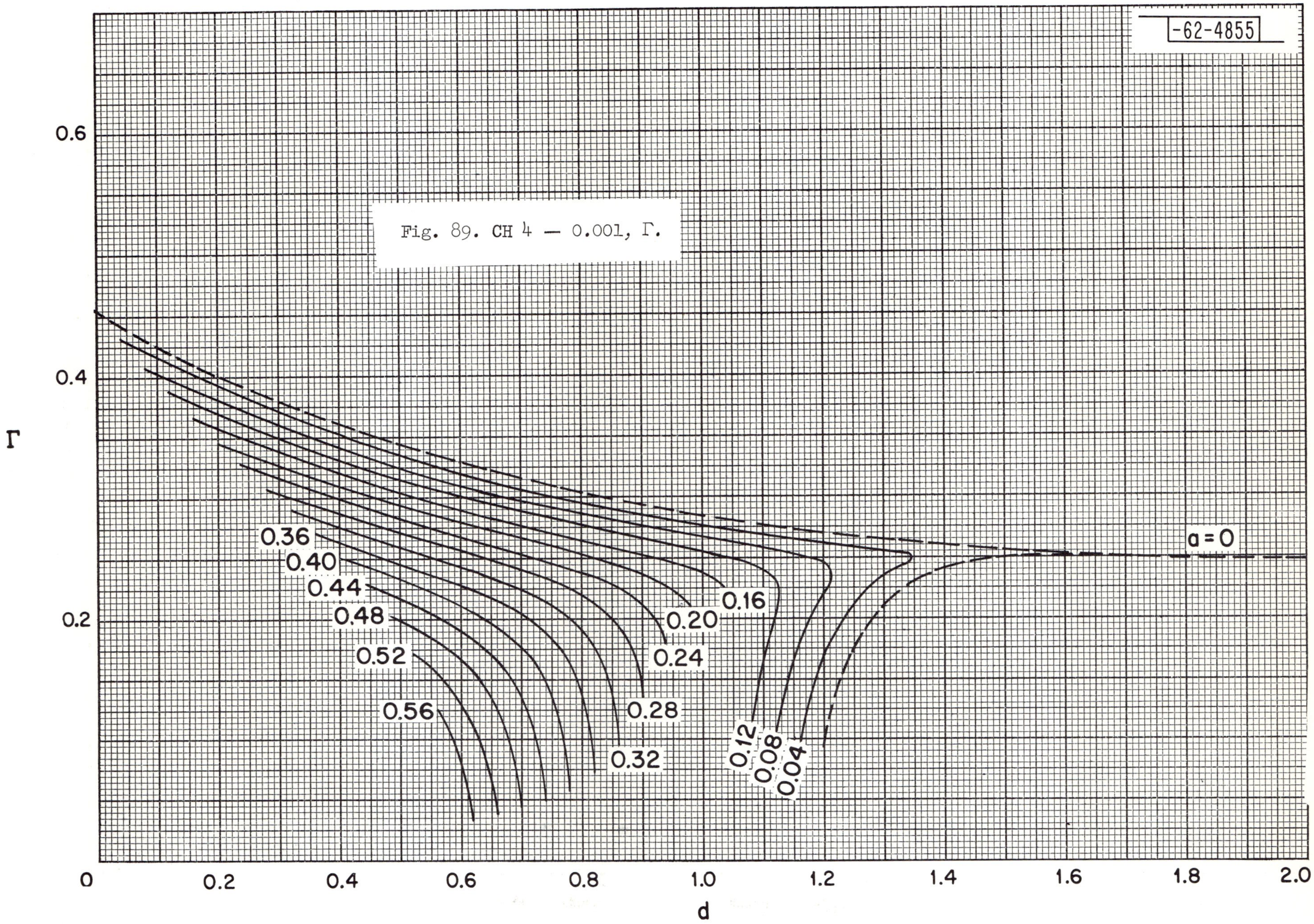


Fig. 89. CH 4 — 0.001, Γ.

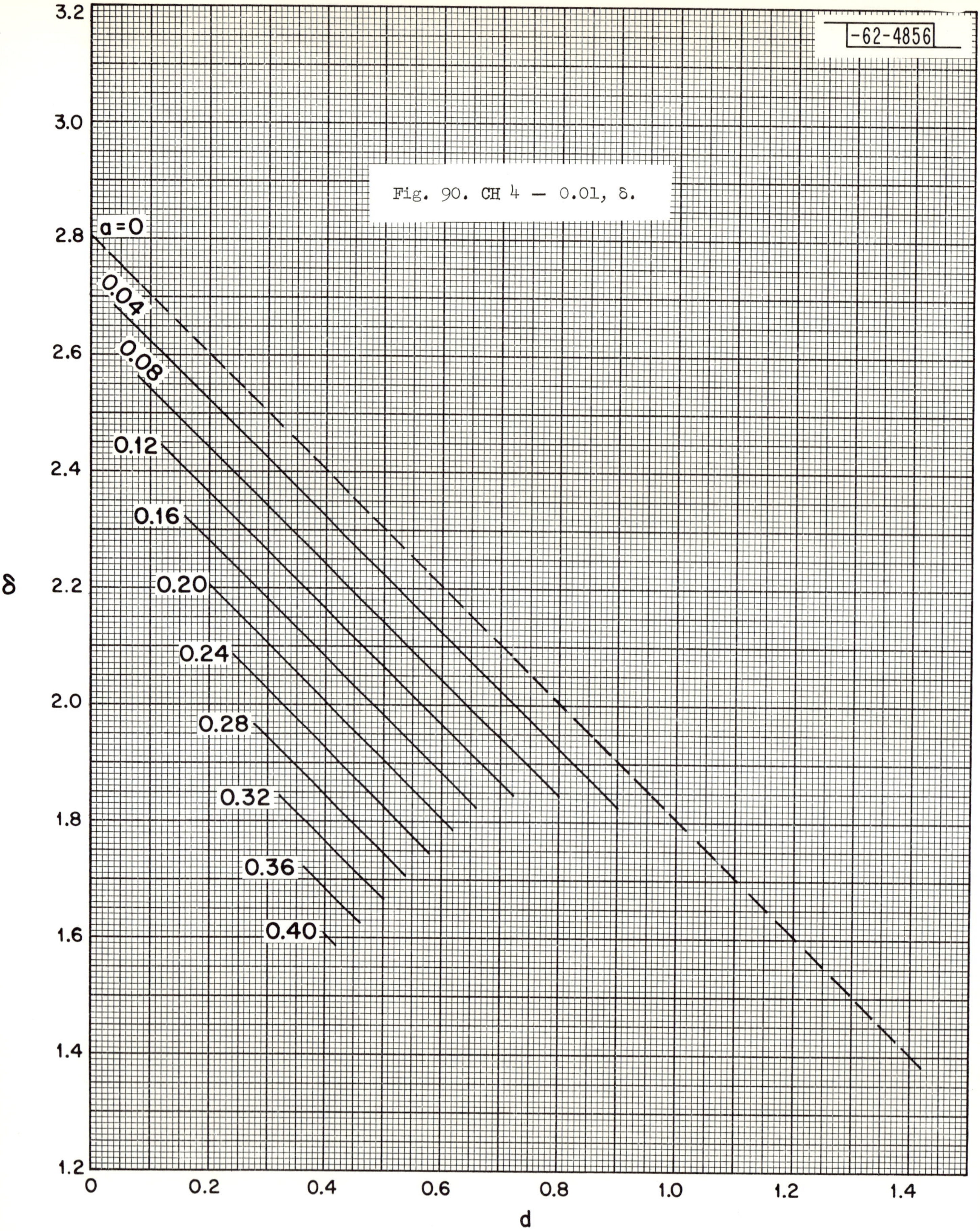


Fig. 90. CH 4 — 0.01, δ.

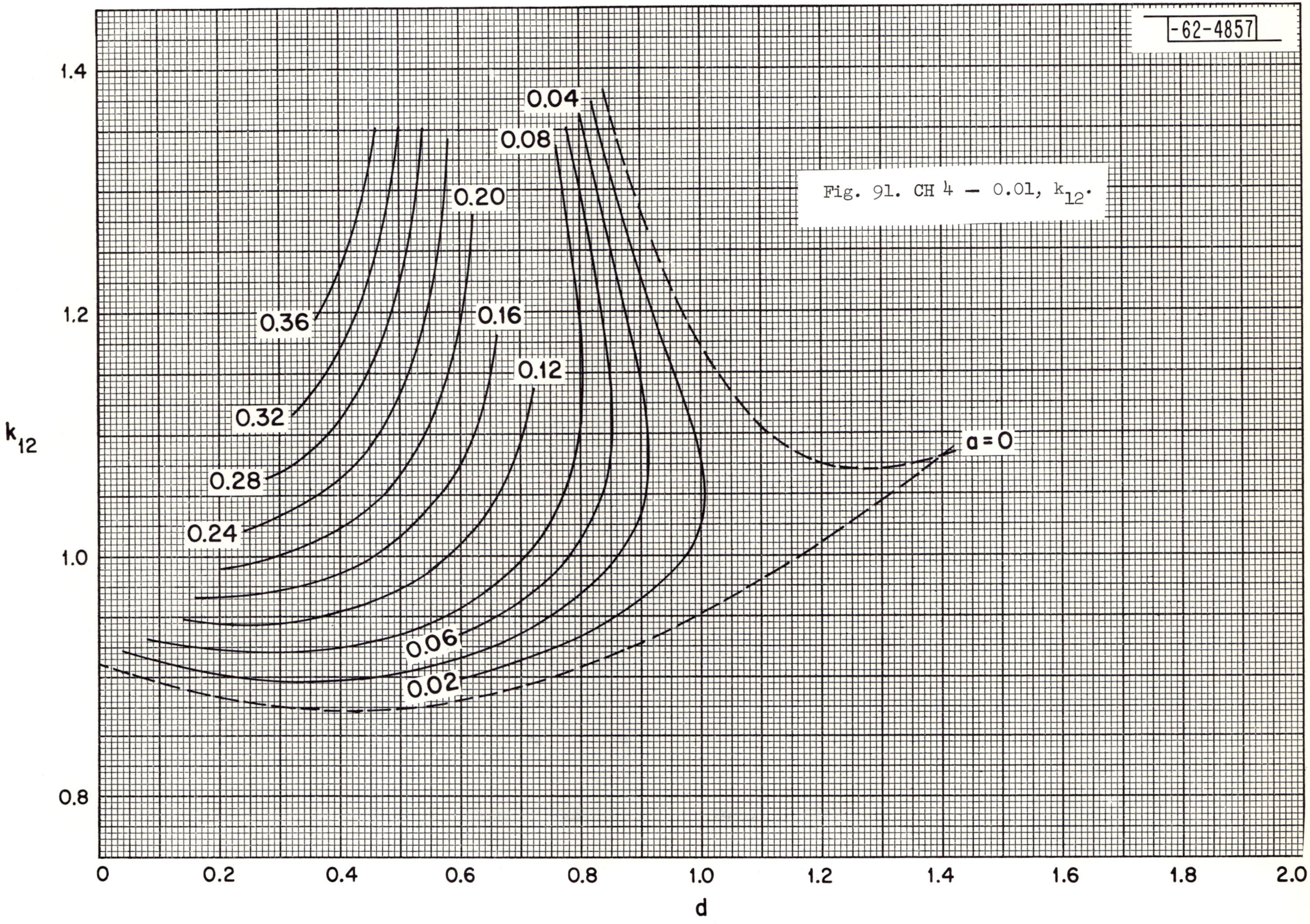


Fig. 91. CH 4 — 0.01, $k_{12}$.

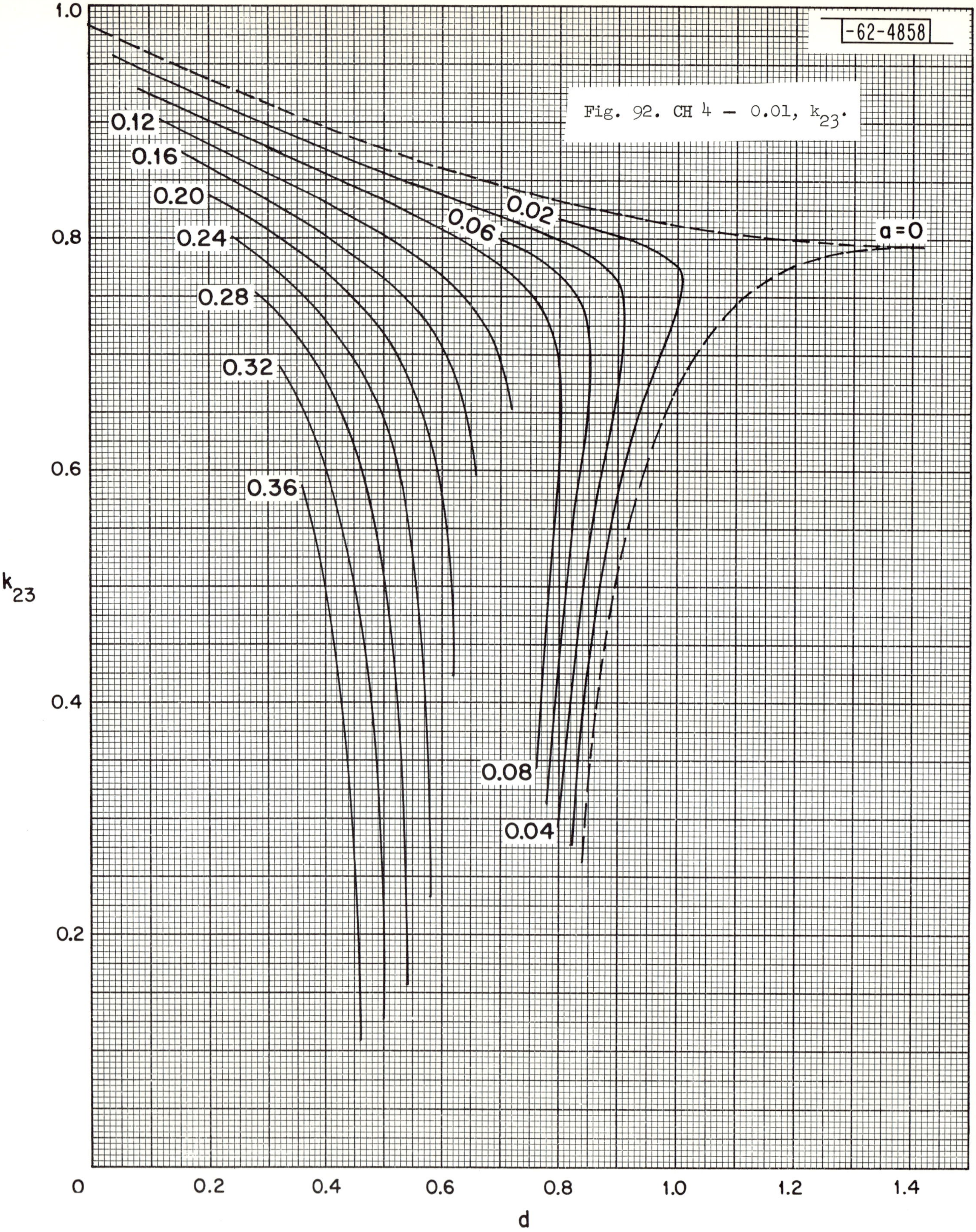


Fig. 92. CH 4 — 0.01, $k_{23}$.

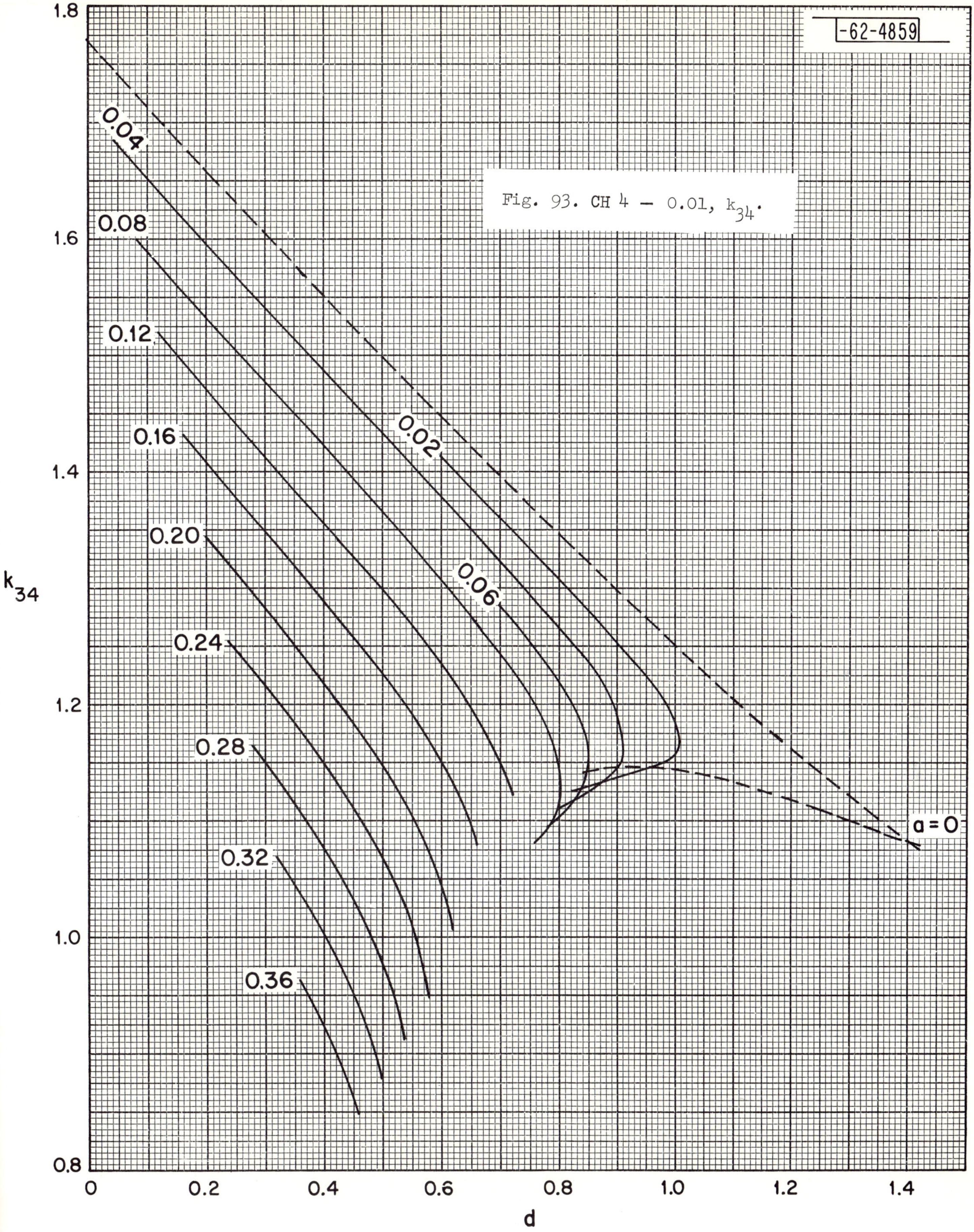


Fig. 93. CH 4 — 0.01, $k_{34}$.

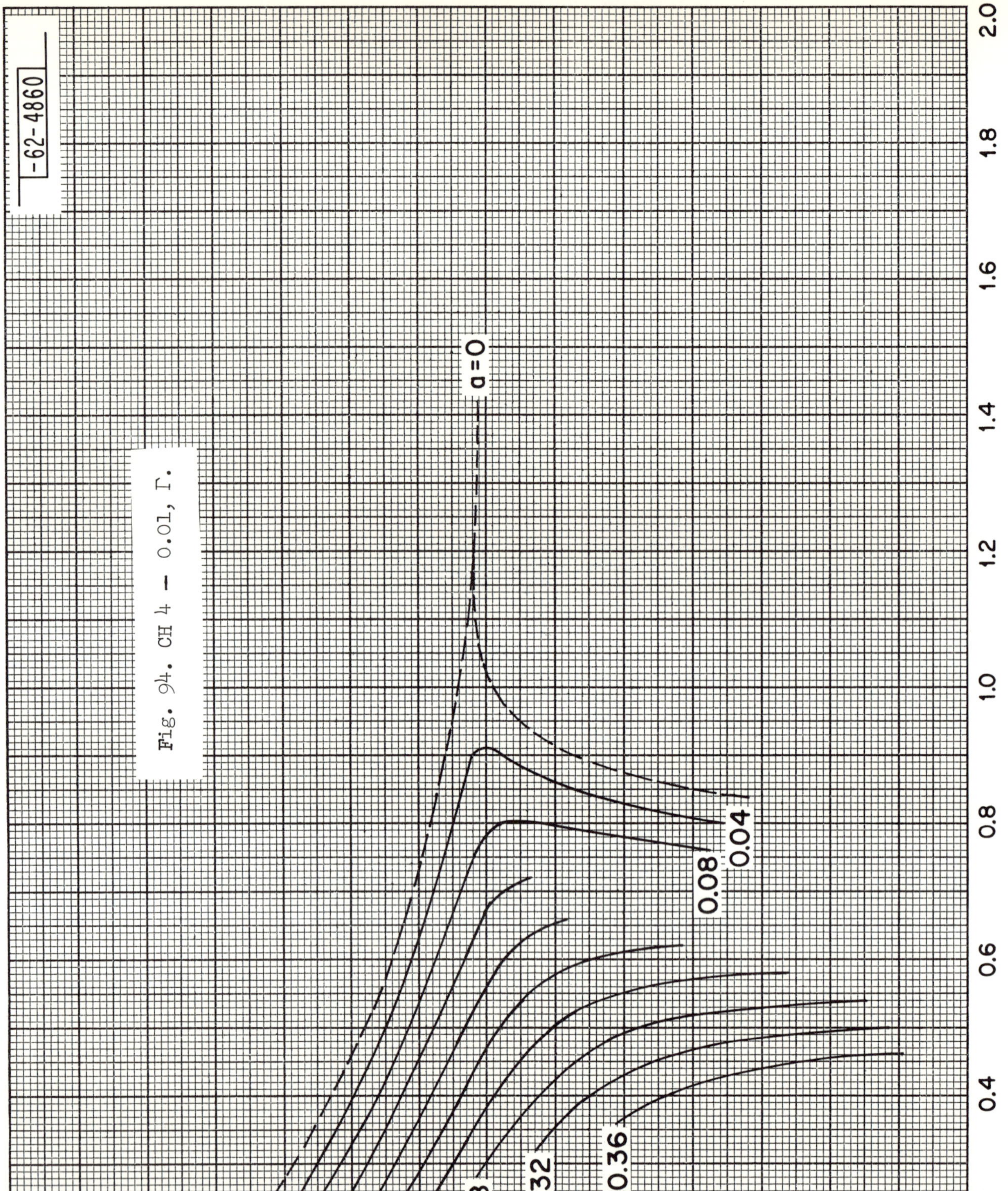

Fig. 94. CH 4 — 0.01, Γ.

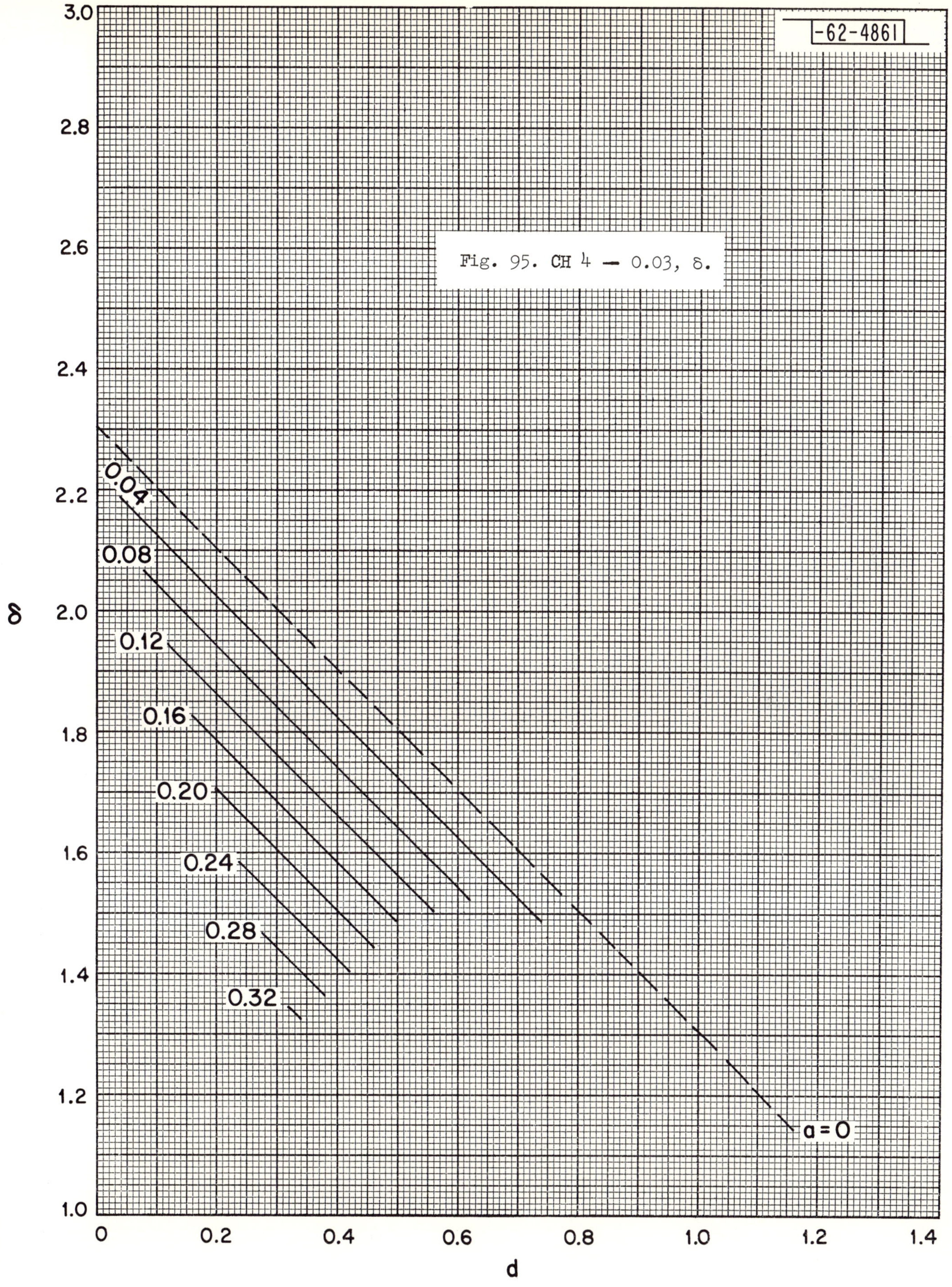
-62-4861
Fig. 95. CH 4 — 0.03, δ.
3.0
2.8
2.6
2.4
2.2
2.0
1.8
1.6
1.4
1.2
1.0
δ
0.04
0.08
0.12
0.16
0.20
0.24
0.28
0.32
a = 0
0
0.2
0.4
0.6
0.8
1.0
1.2
1.4
d

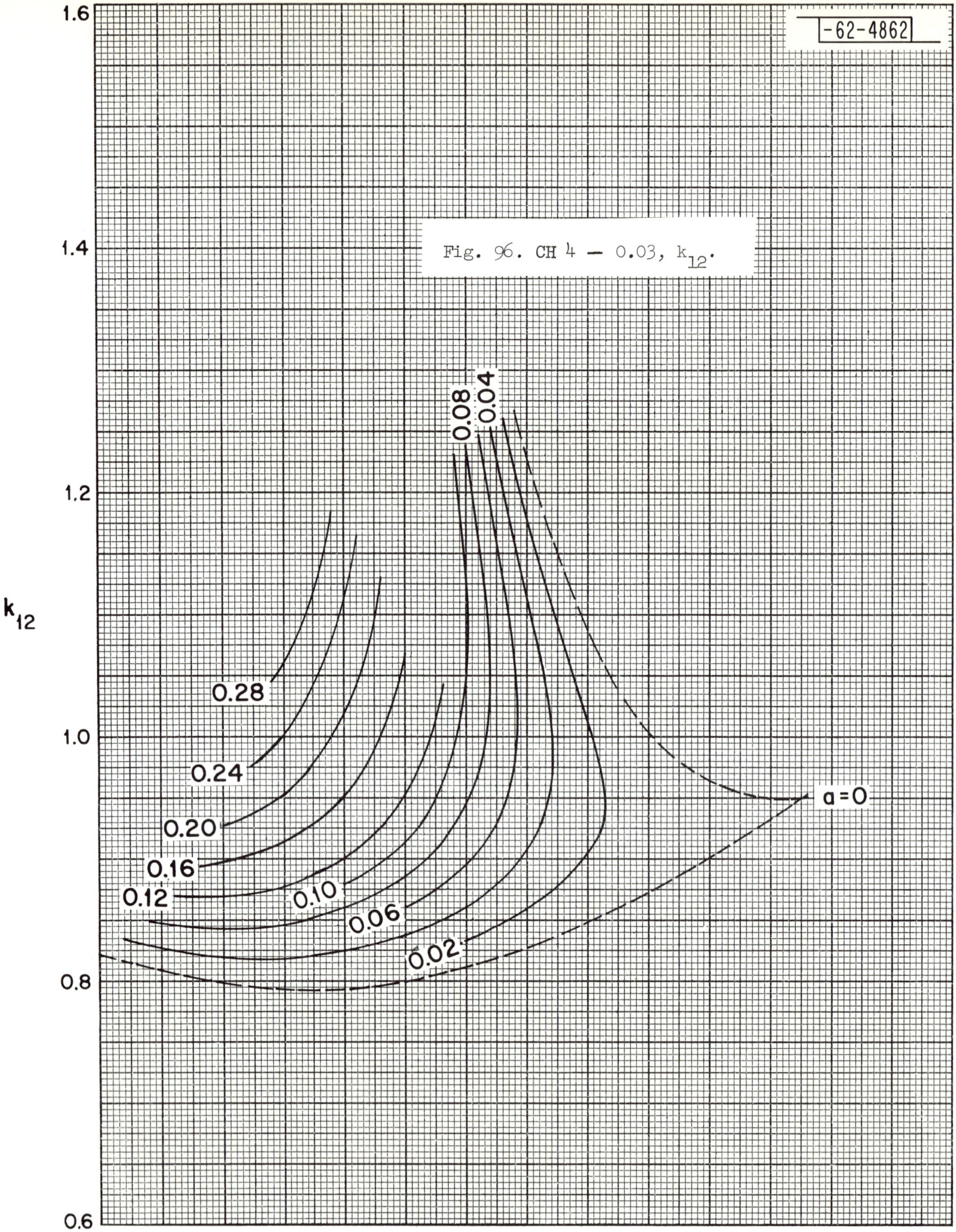
-62-4862
Fig. 96. CH 4 — 0.03, k12.
k12
d
1.6
1.4
1.2
1.0
0.8
0.6
0
0.2
0.4
0.6
0.8
1.0
1.2
1.4
0.28
0.24
0.20
0.16
0.12
0.10
0.08
0.06
0.04
0.02
a=0

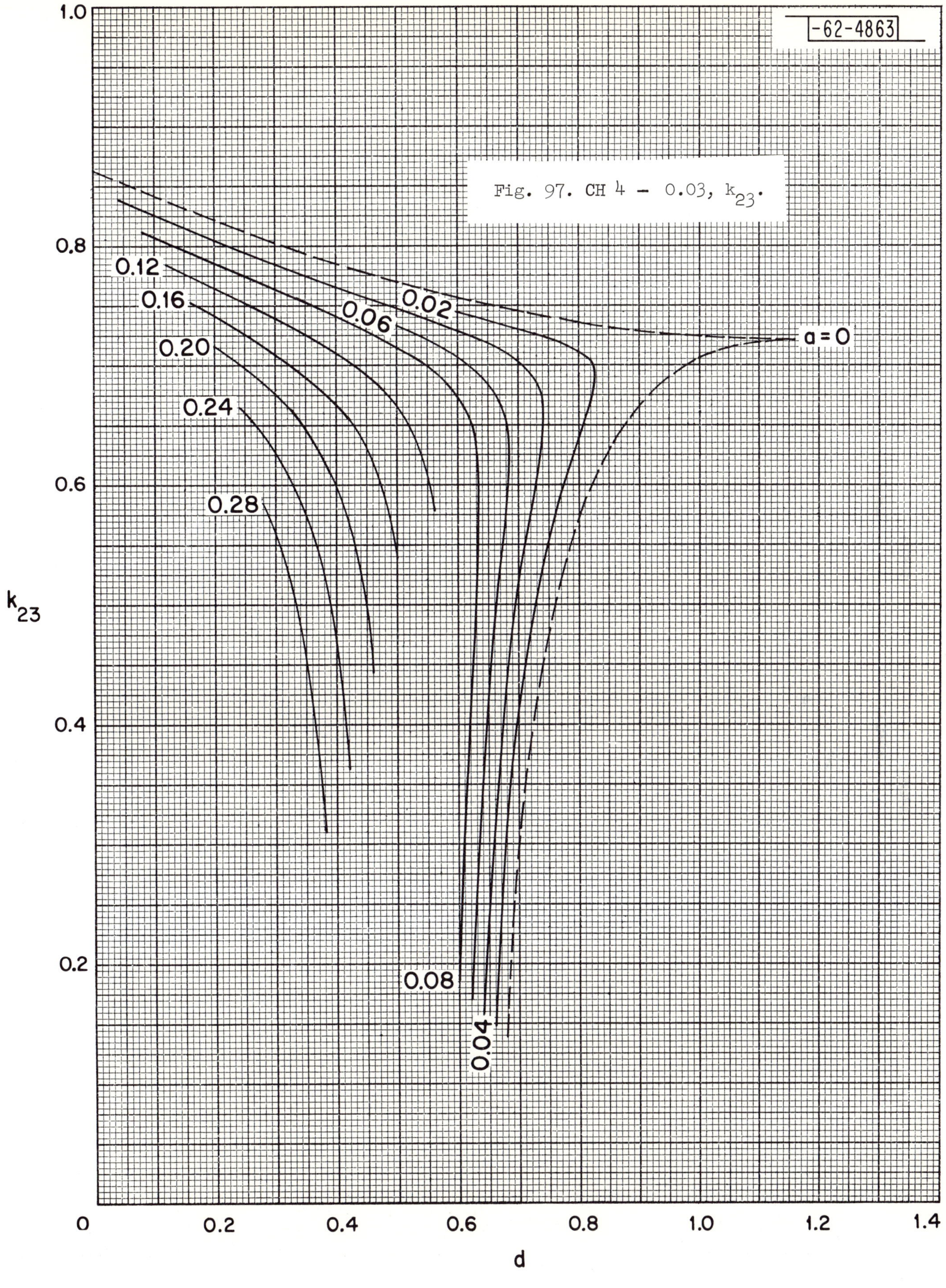
-62-4863
Fig. 97. CH 4 — 0.03, k23.
1.0
0.8
0.6
0.4
0.2
k23
0
0.2
0.4
0.6
0.8
1.0
1.2
1.4
d
0.02
0.06
0.12
0.16
0.20
0.24
0.28
0.08
0.04
a = 0

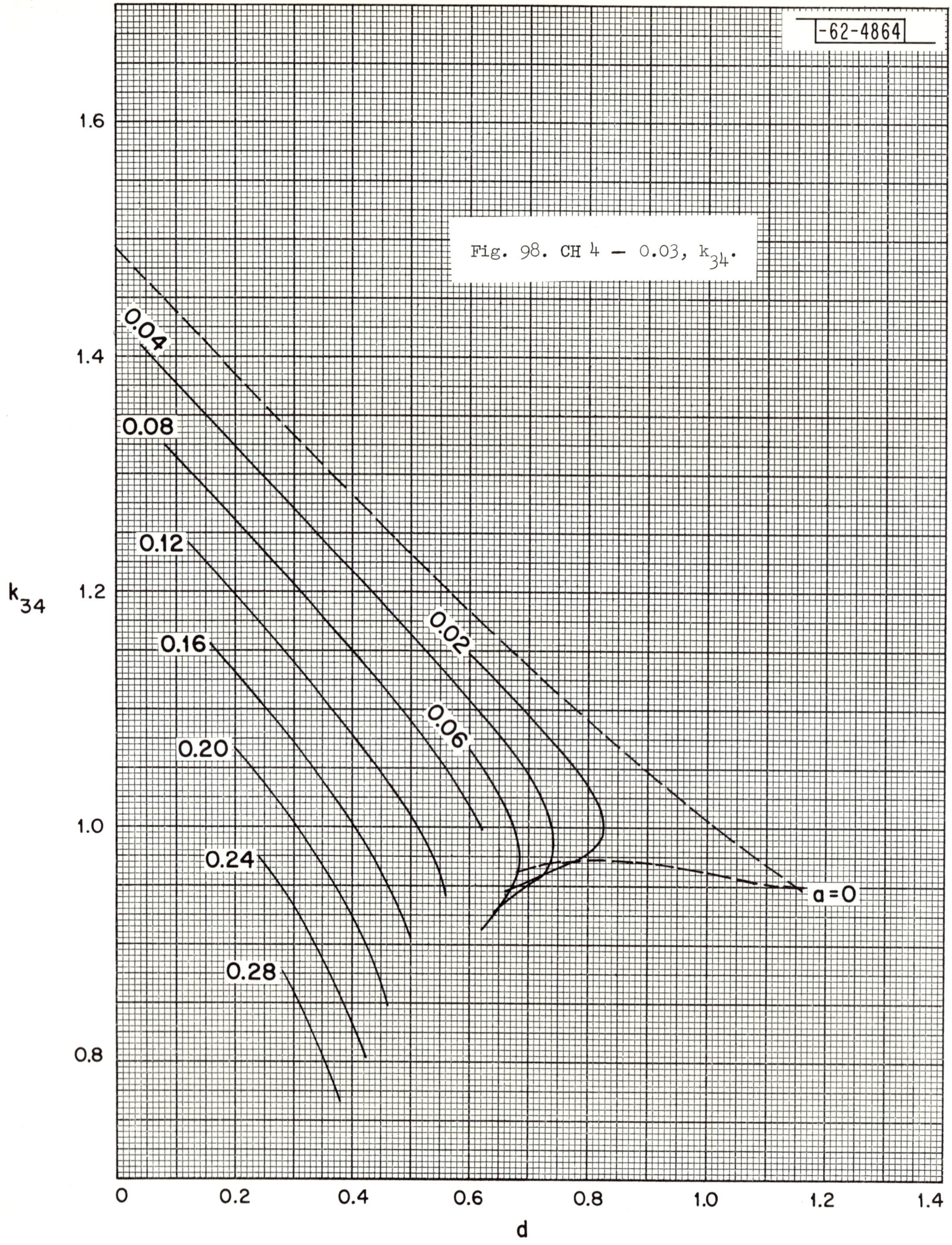
-62-4864
Fig. 98. CH 4 — 0.03, $k_{34}$.
1.6
1.4
1.2
1.0
0.8
$k_{34}$
0.04
0.08
0.12
0.16
0.20
0.24
0.28
0.02
0.06
a=0
0
0.2
0.4
0.6
0.8
1.0
1.2
1.4
d

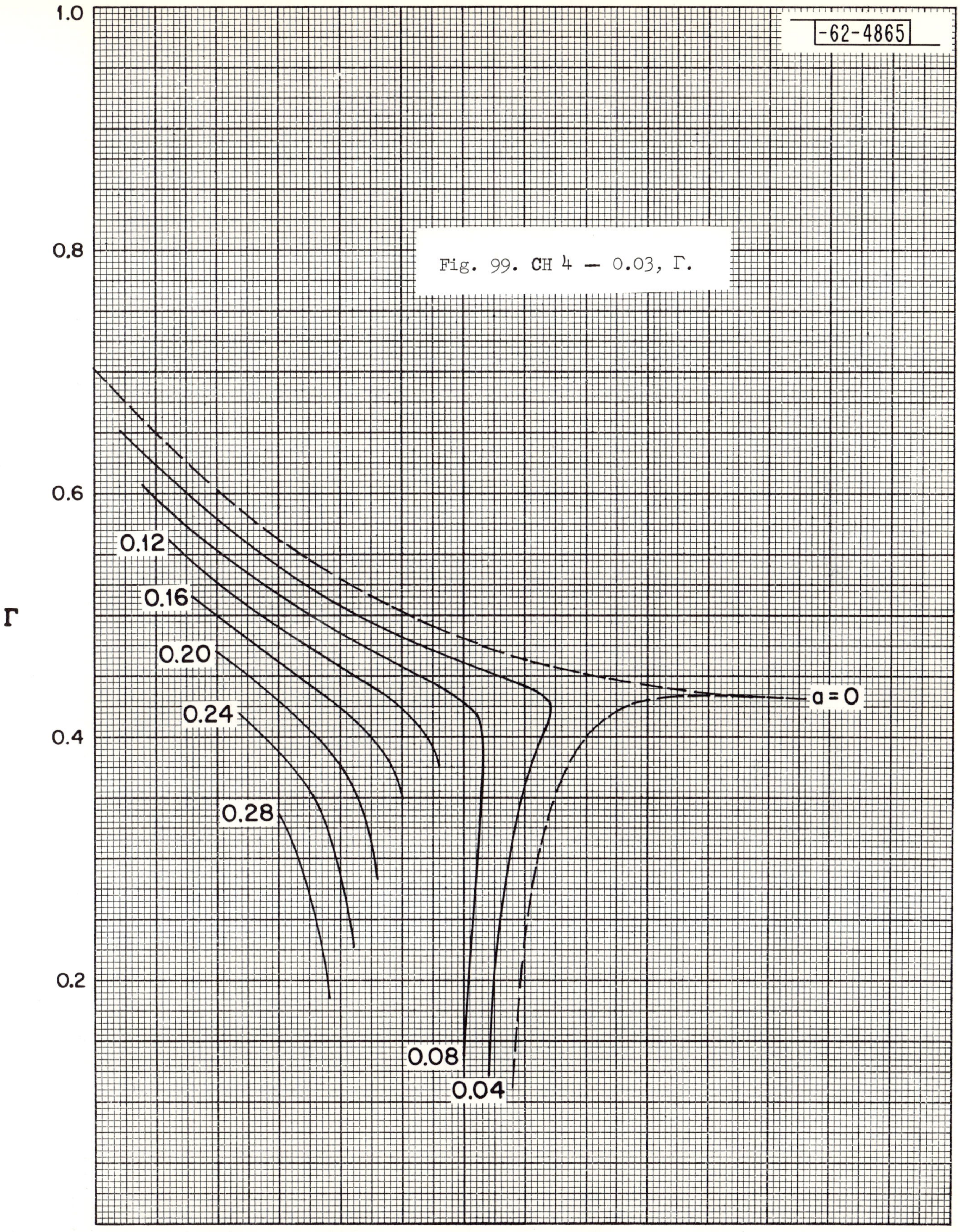
-62-4865
Fig. 99. CH 4 — 0.03, Γ.
1.0
0.8
0.6
0.4
0.2
Γ
0
0.2
0.4
0.6
0.8
1.0
1.2
1.4
d
0.12
0.16
0.20
0.24
0.28
0.08
0.04
a = 0

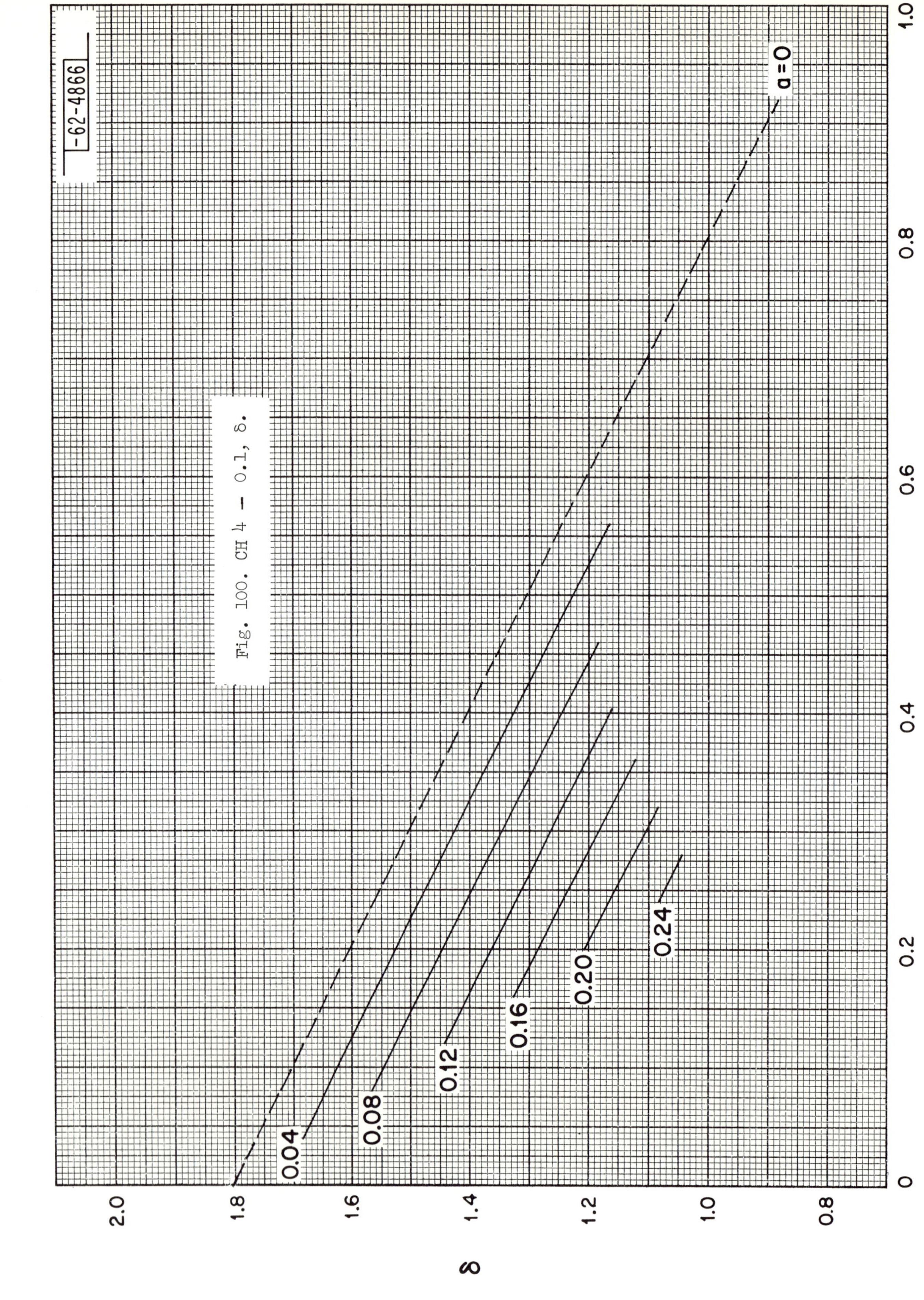

Fig. 100. CH 4 — 0.1, δ.

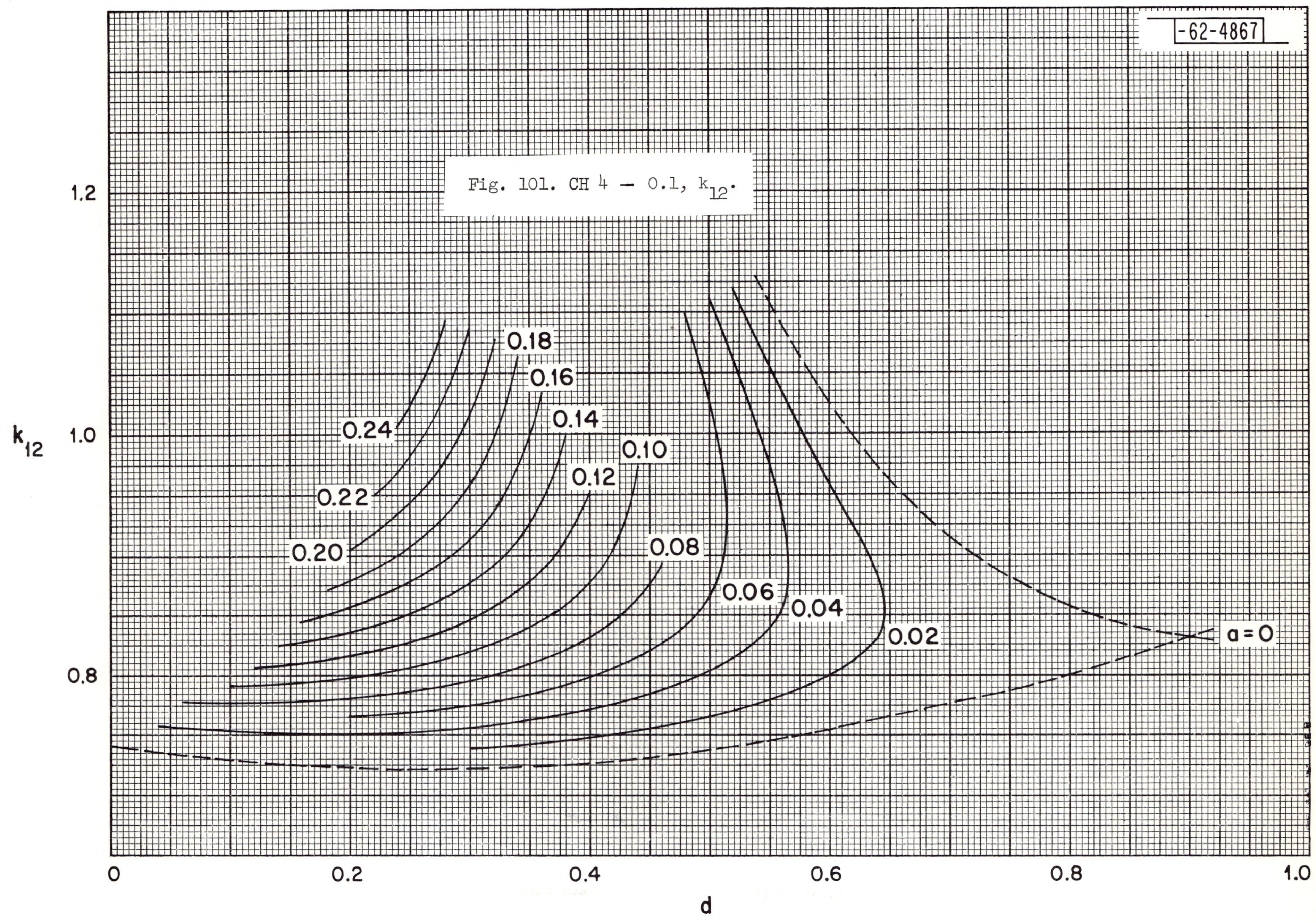


Fig. 101. CH 4 — 0.1, $k_{12}$.

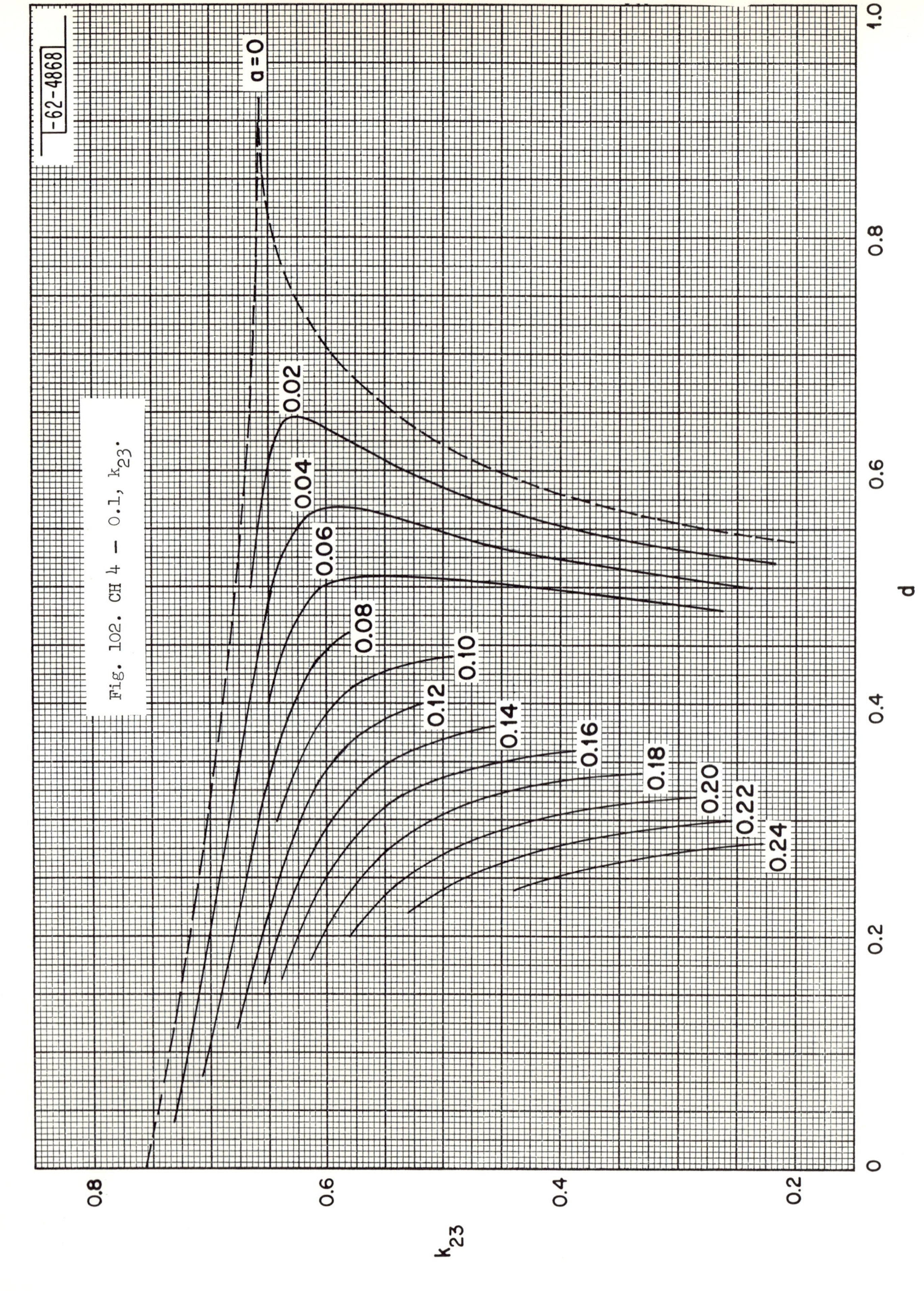


Fig. 102. CH 4 — 0.1, $k_{23}$.

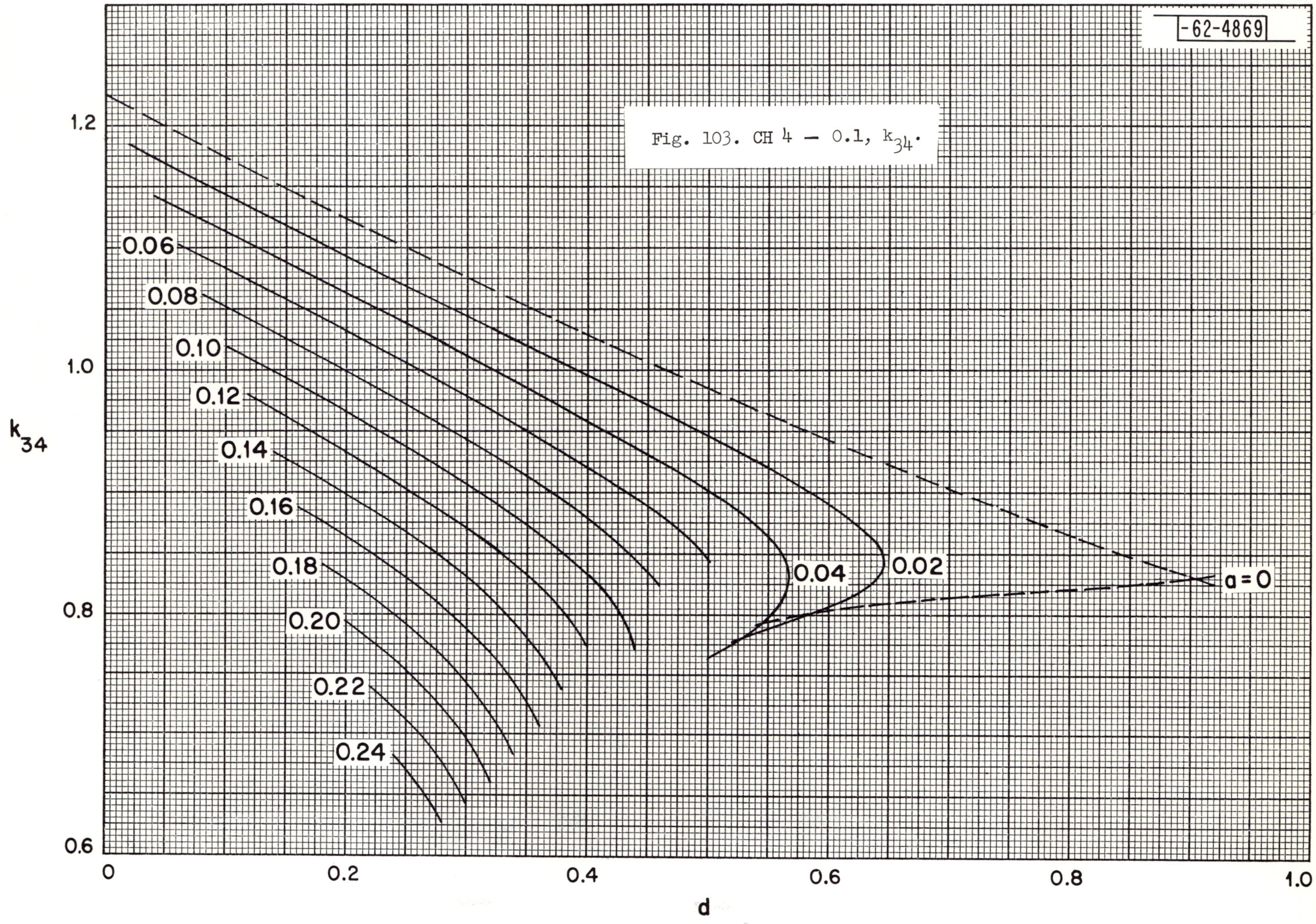


Fig. 103. CH 4 — 0.1, $k_{34}$.

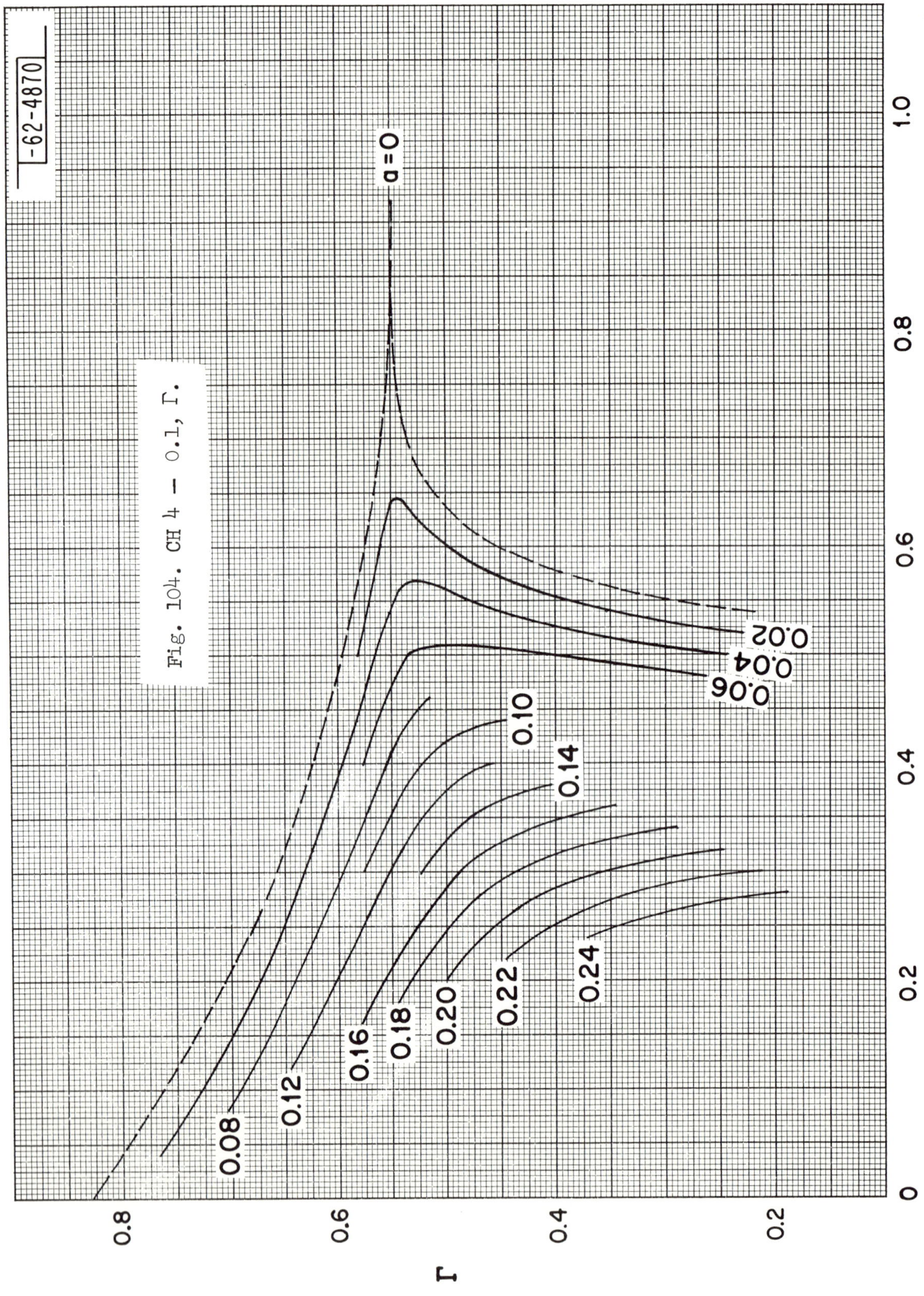


Fig. 104. CH 4 — 0.1, Γ.

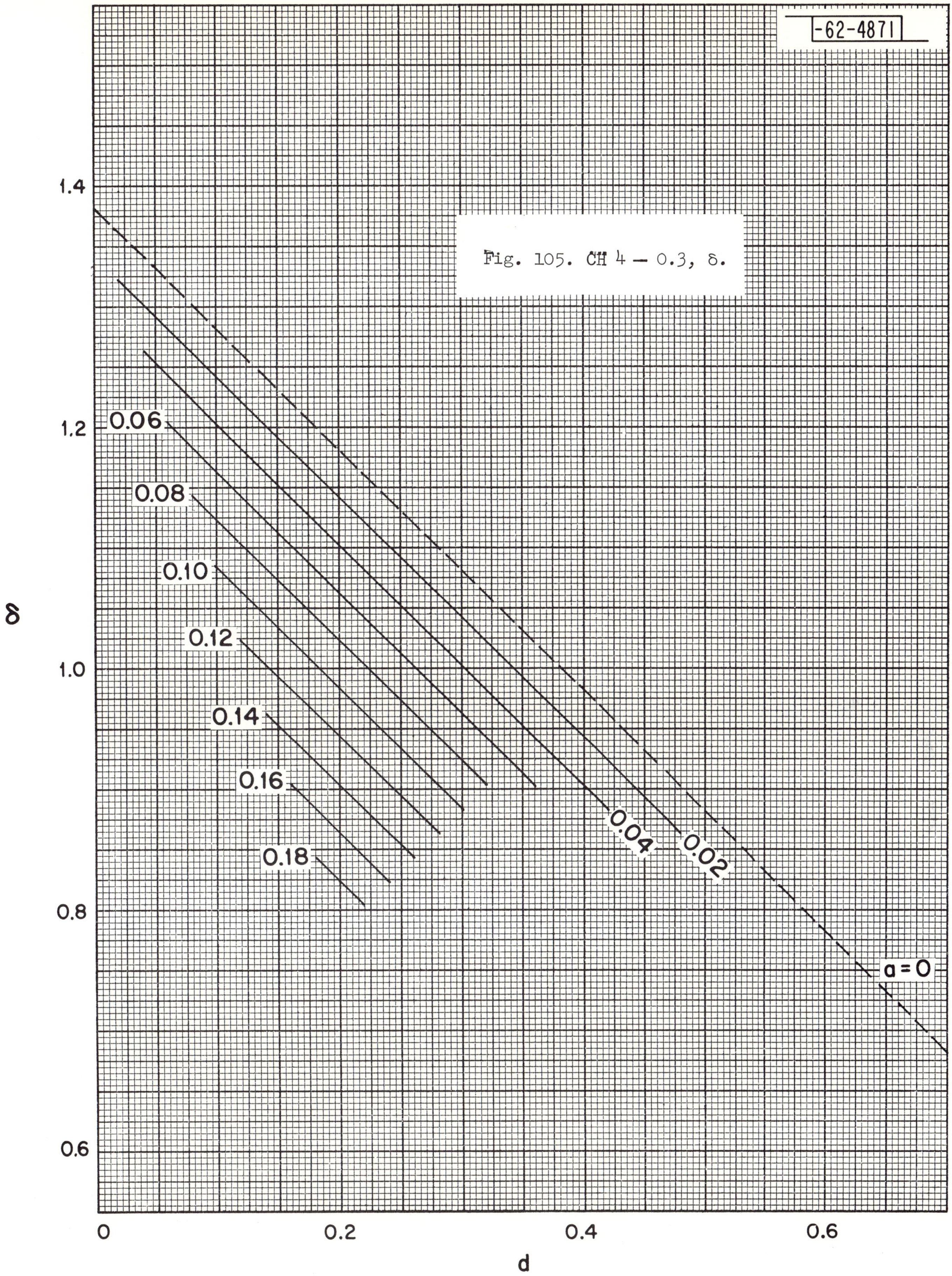


Fig. 105. CH 4 — 0.3, δ.

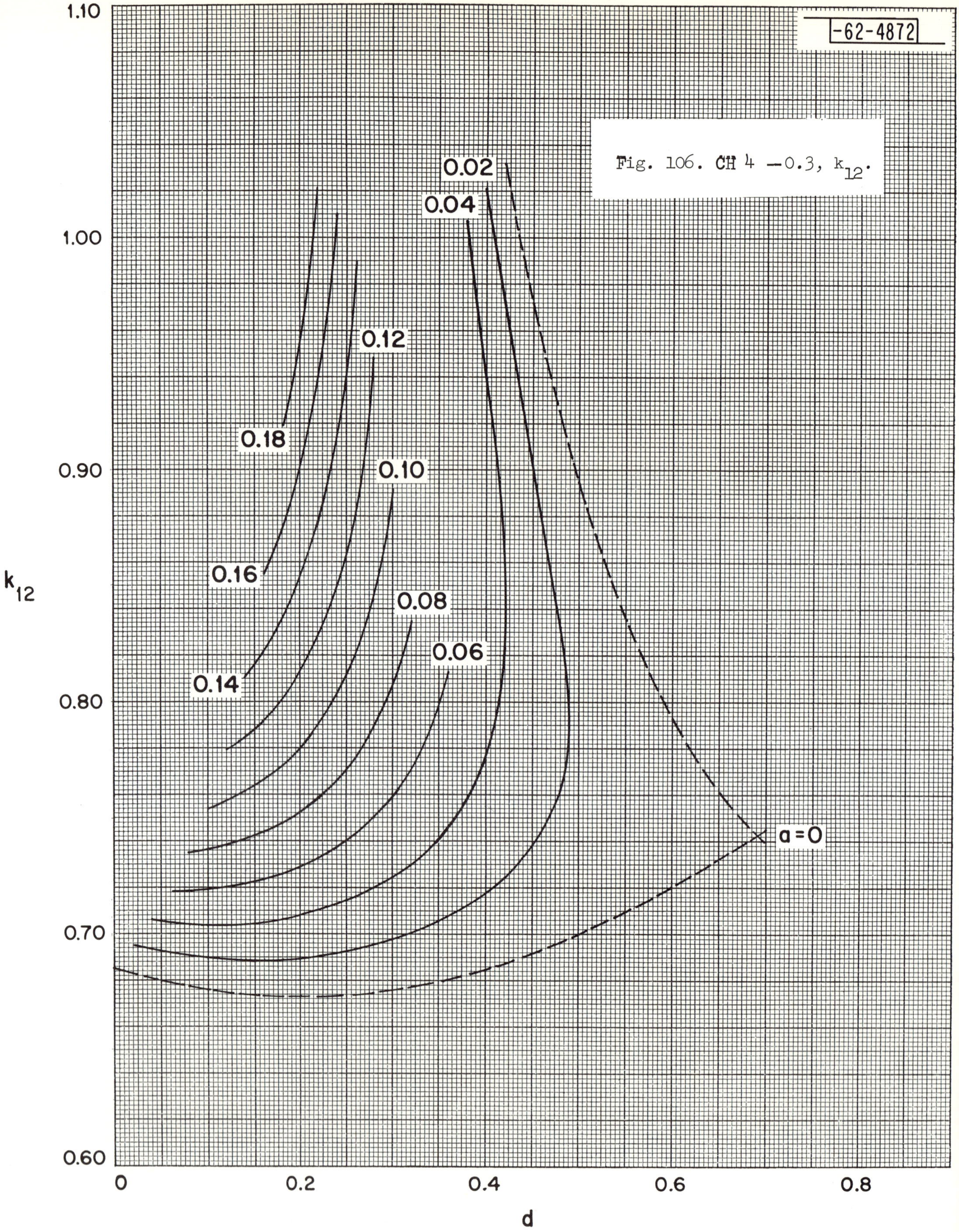


Fig. 106. CH 4 —0.3, $k_{12}$.

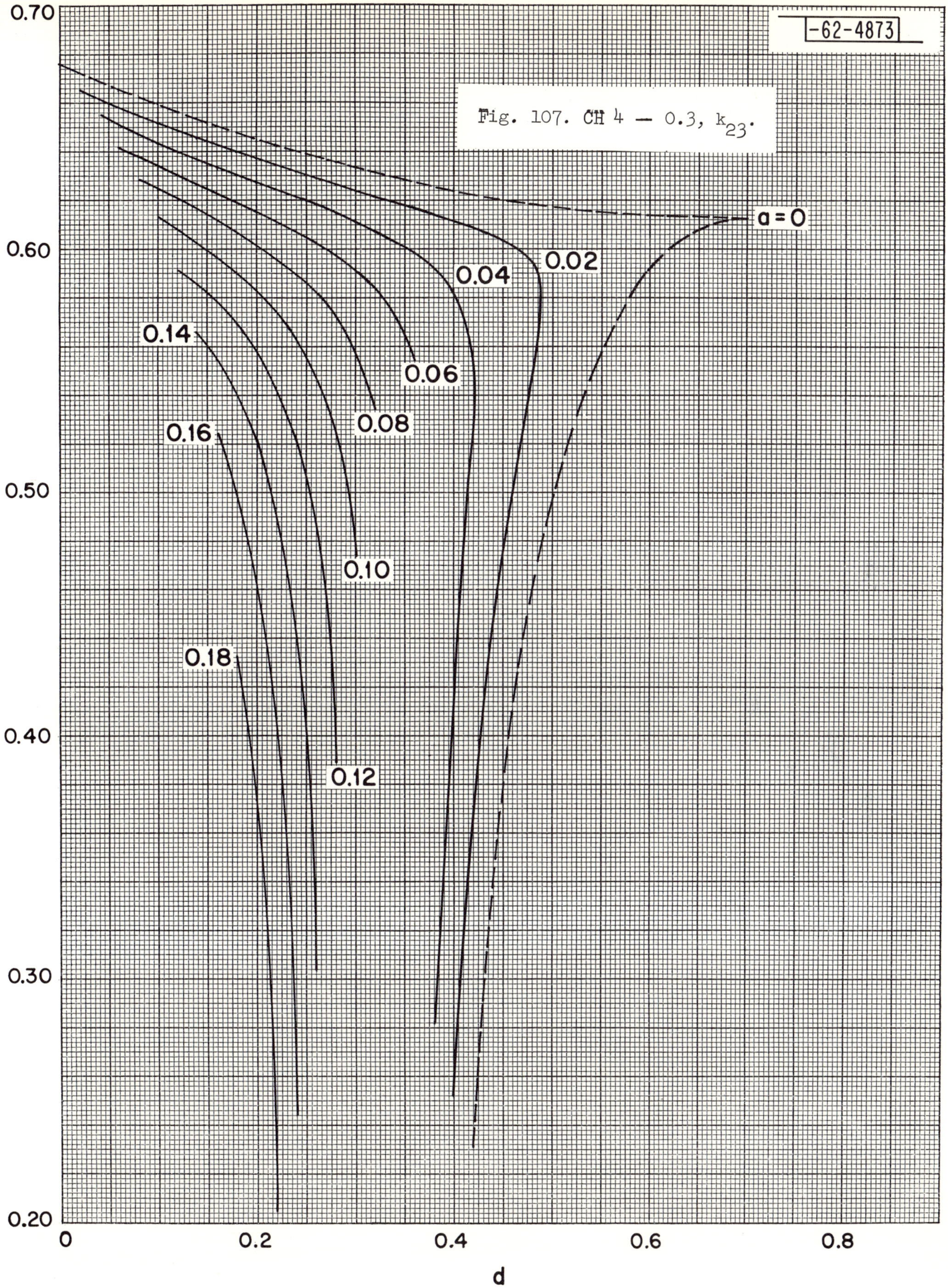


Fig. 107. CH 4 — 0.3, $k_{23}$.

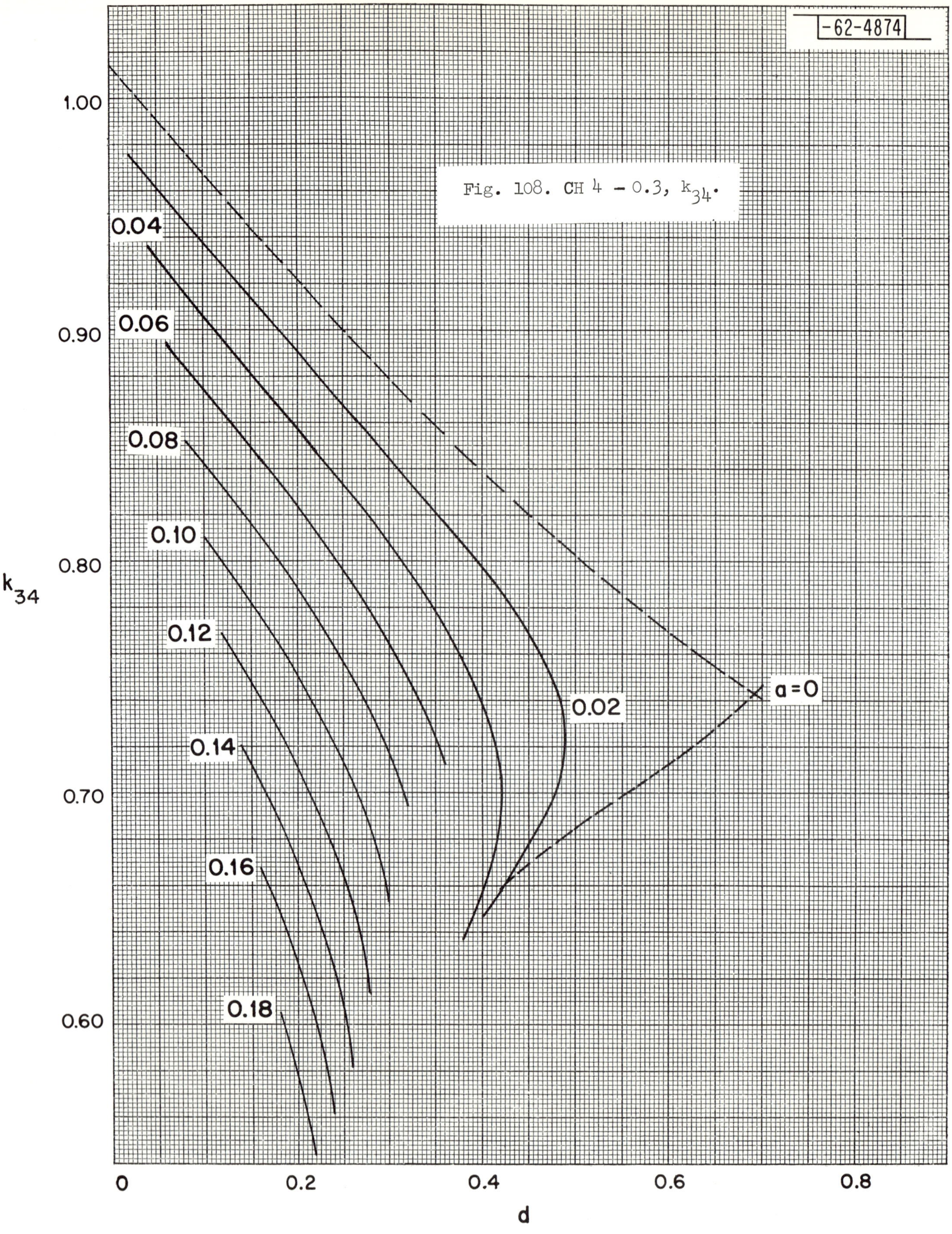


Fig. 108. CH 4 – 0.3, $k_{34}$.

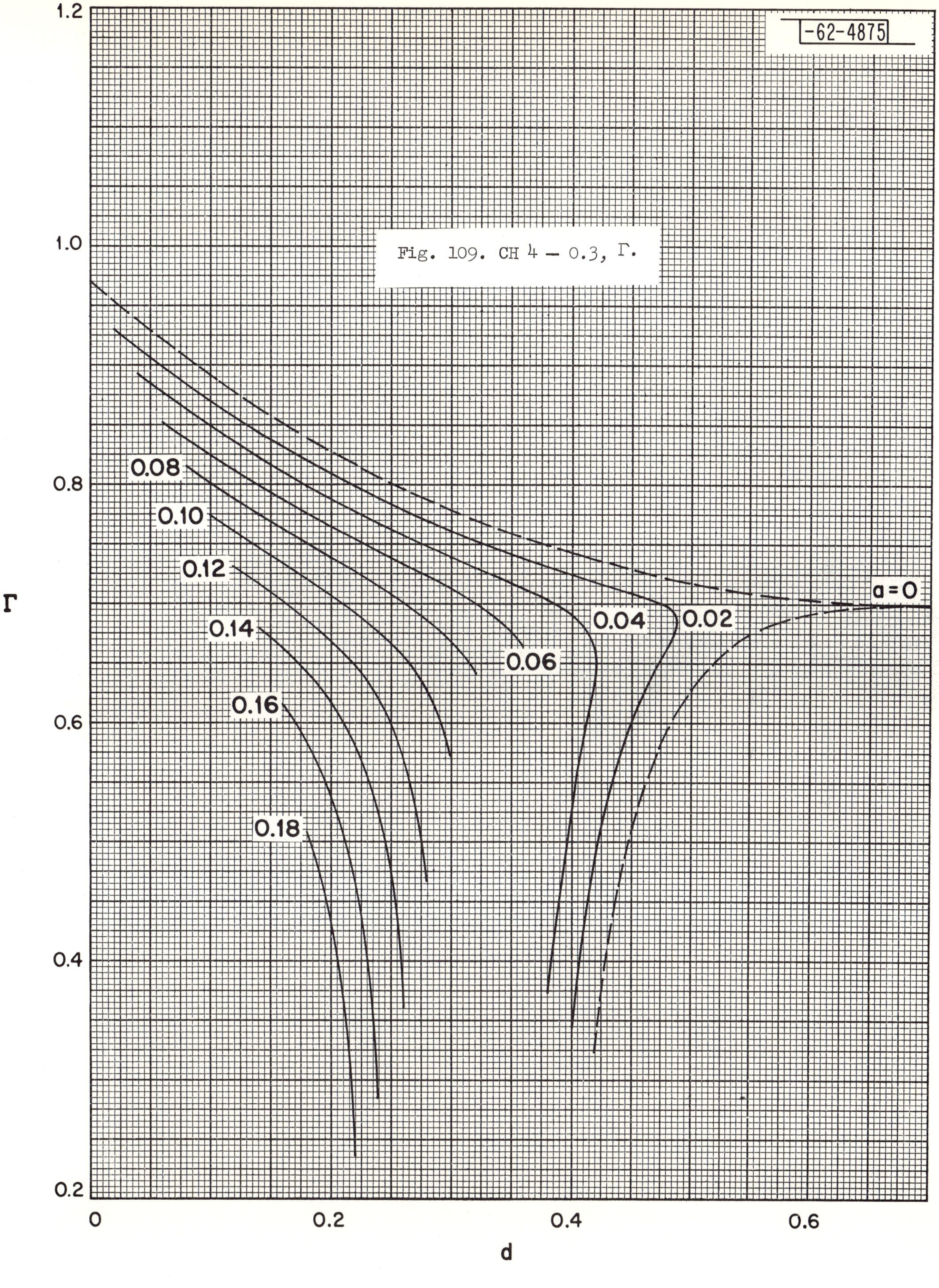

L-62-4875
Fig. 109. CH 4 — 0.3, Γ.
Γ
d
1.2
1.0
0.8
0.6
0.4
0.2
0
0.2
0.4
0.6
0.08
0.10
0.12
0.14
0.16
0.18
0.06
0.04
0.02
a=0

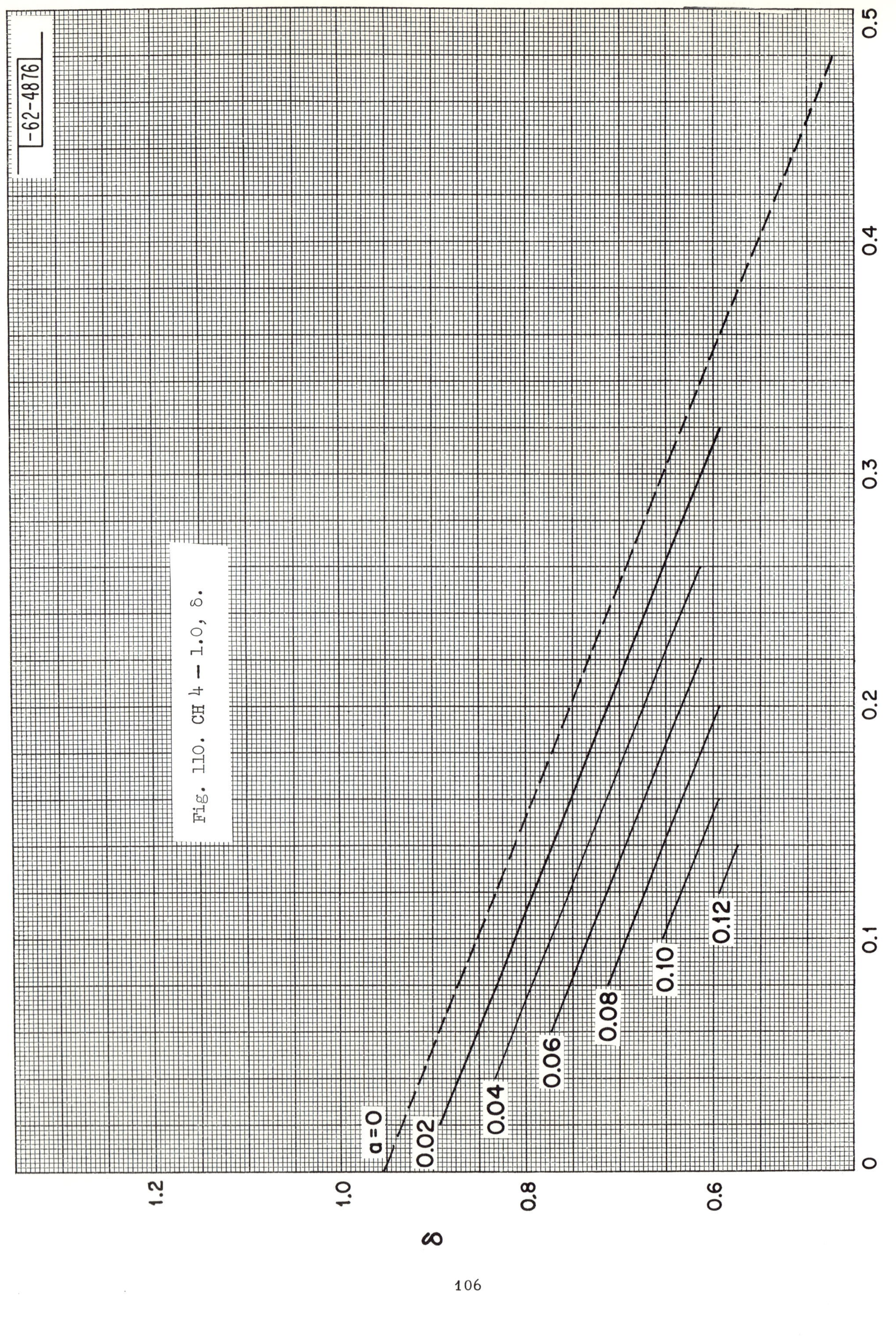


Fig. 110. CH 4 — 1.0, δ.

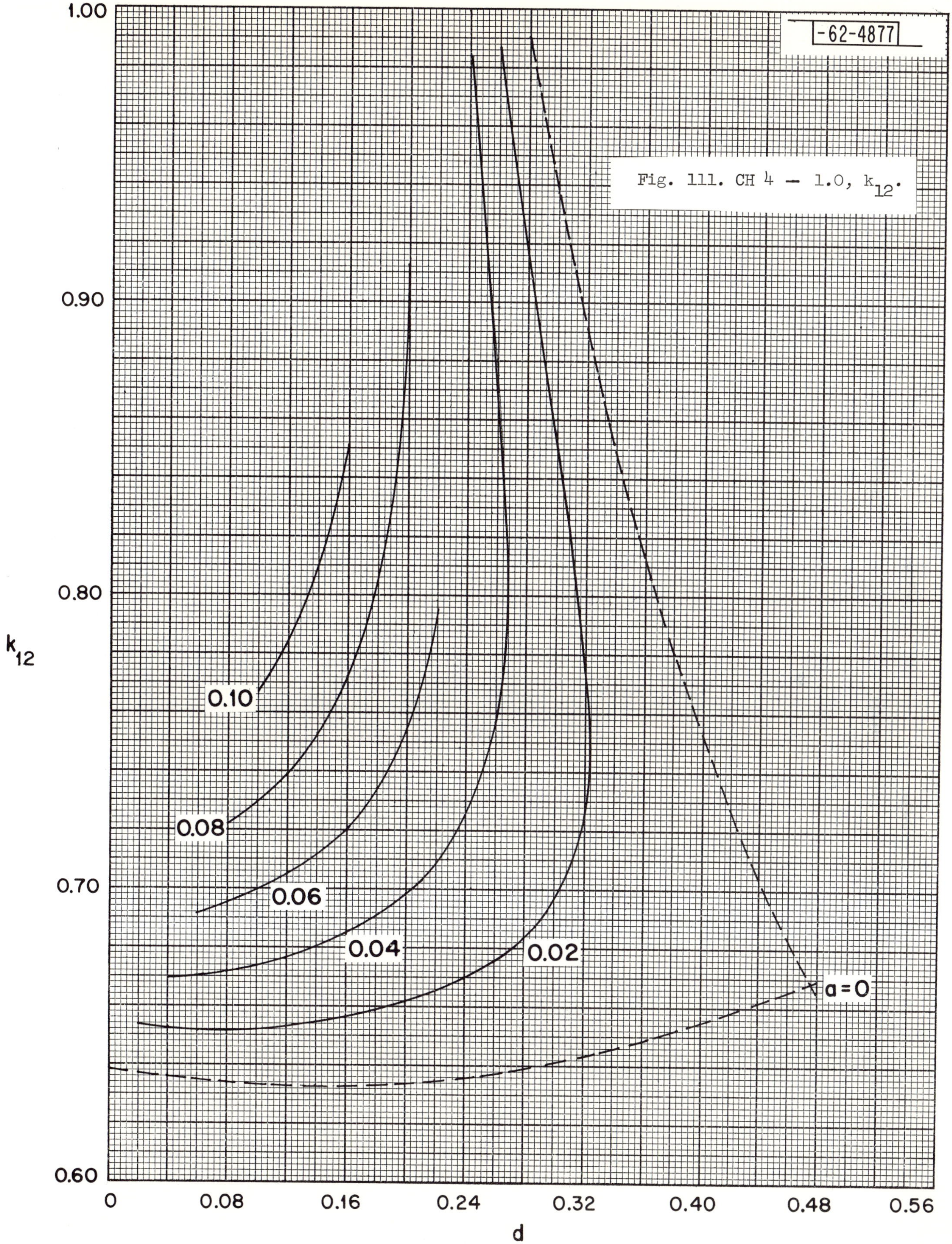


Fig. 111. $CH_4$ — 1.0, $k_{12}$.

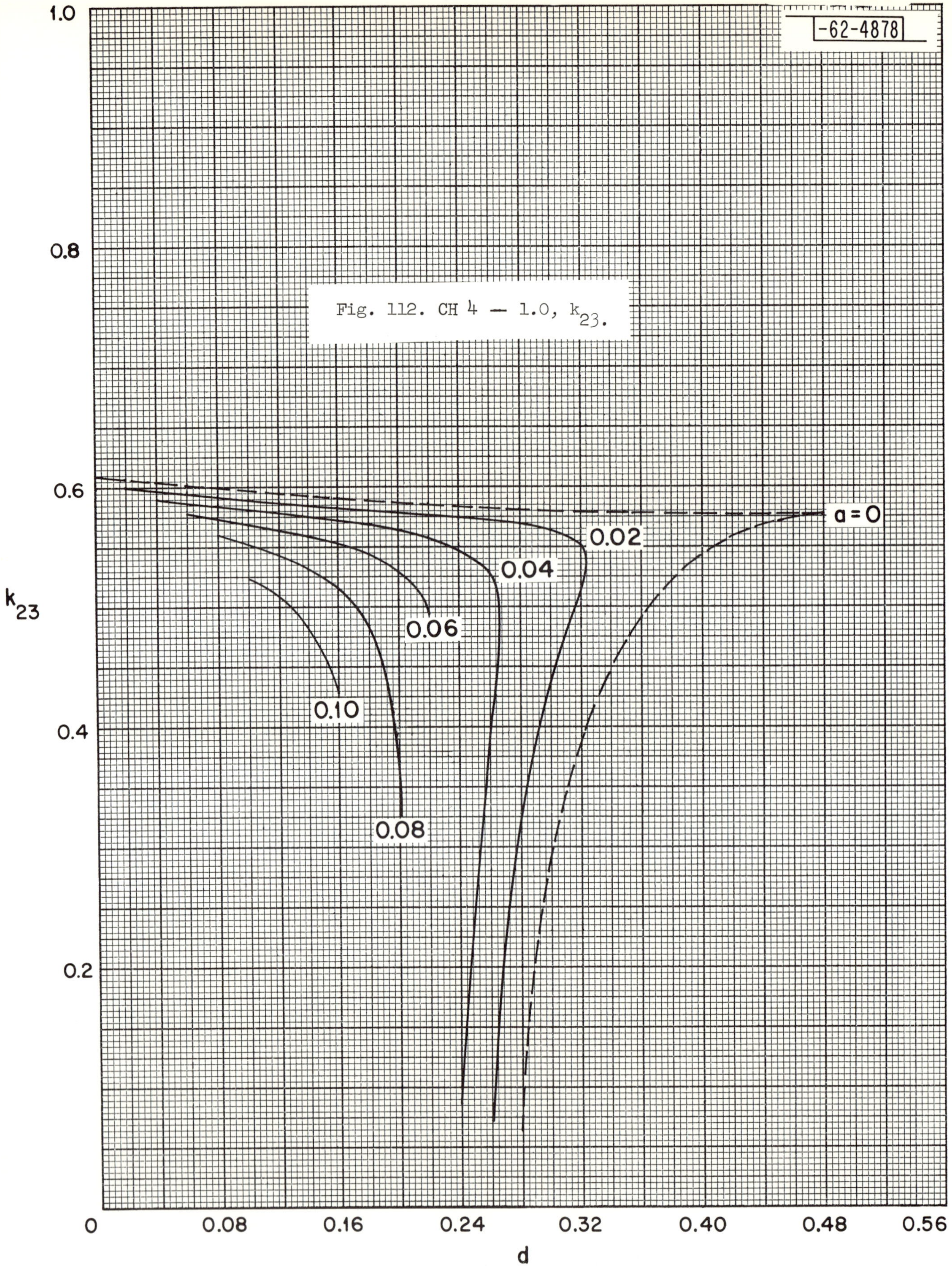
-62-4878
Fig. 112. CH 4 — 1.0, $k_{23}$.
1.0
0.8
0.6
0.4
0.2
$k_{23}$
a = 0
0.02
0.04
0.06
0.08
0.10
0
0.08
0.16
0.24
0.32
0.40
0.48
0.56
d

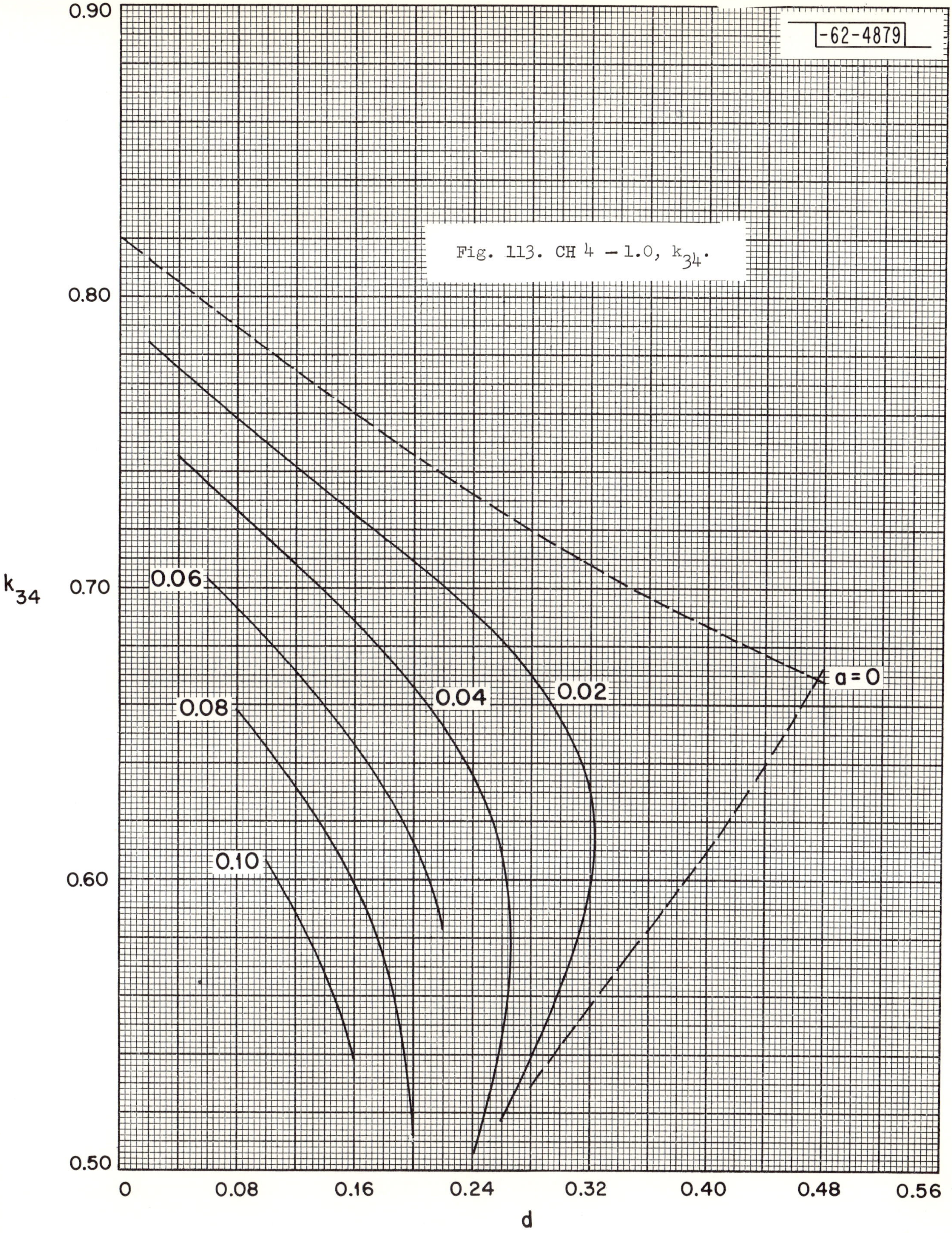


Fig. 113. CH 4 — 1.0, $k_{34}$.

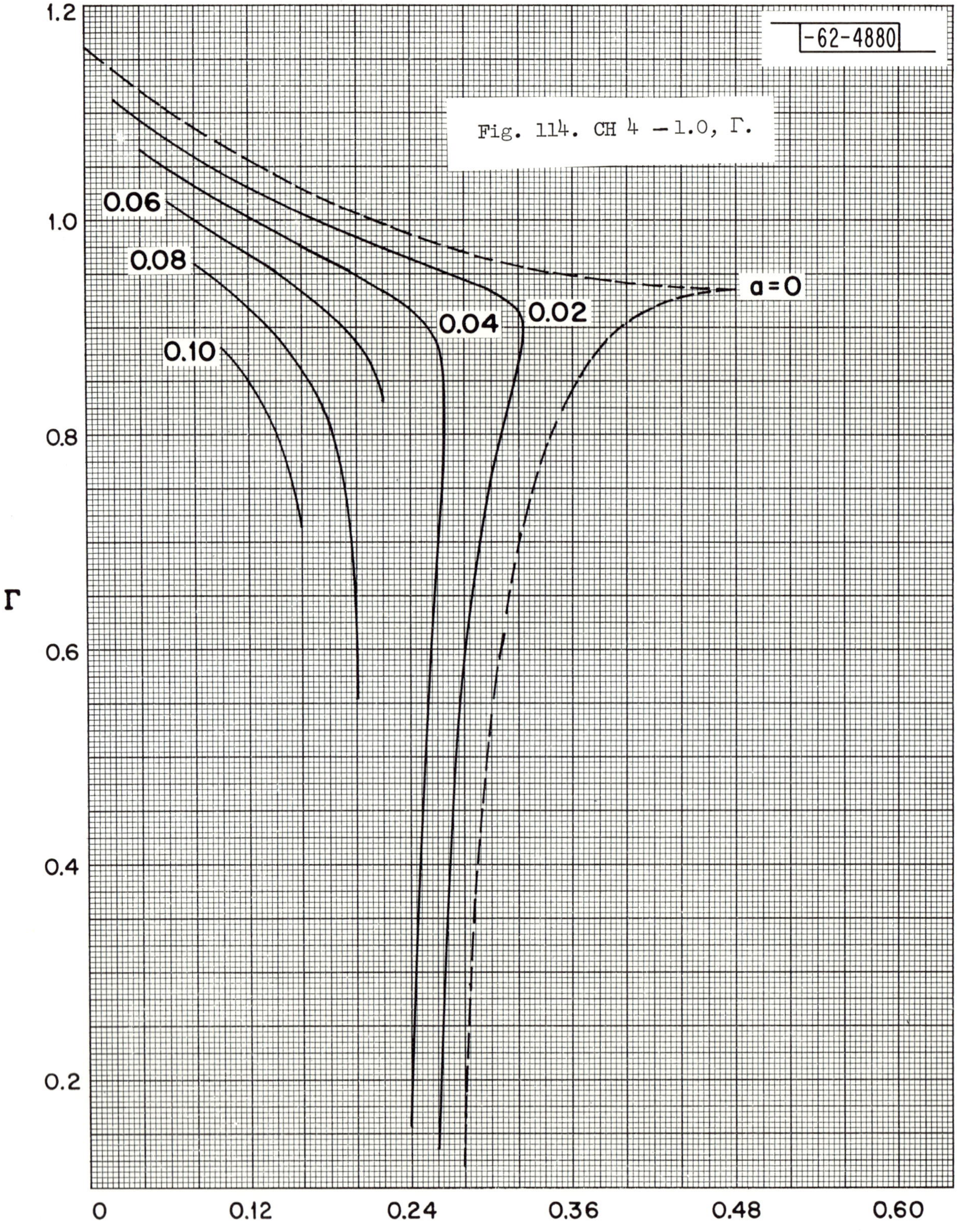


Fig. 114. CH 4 — 1.0, Γ.

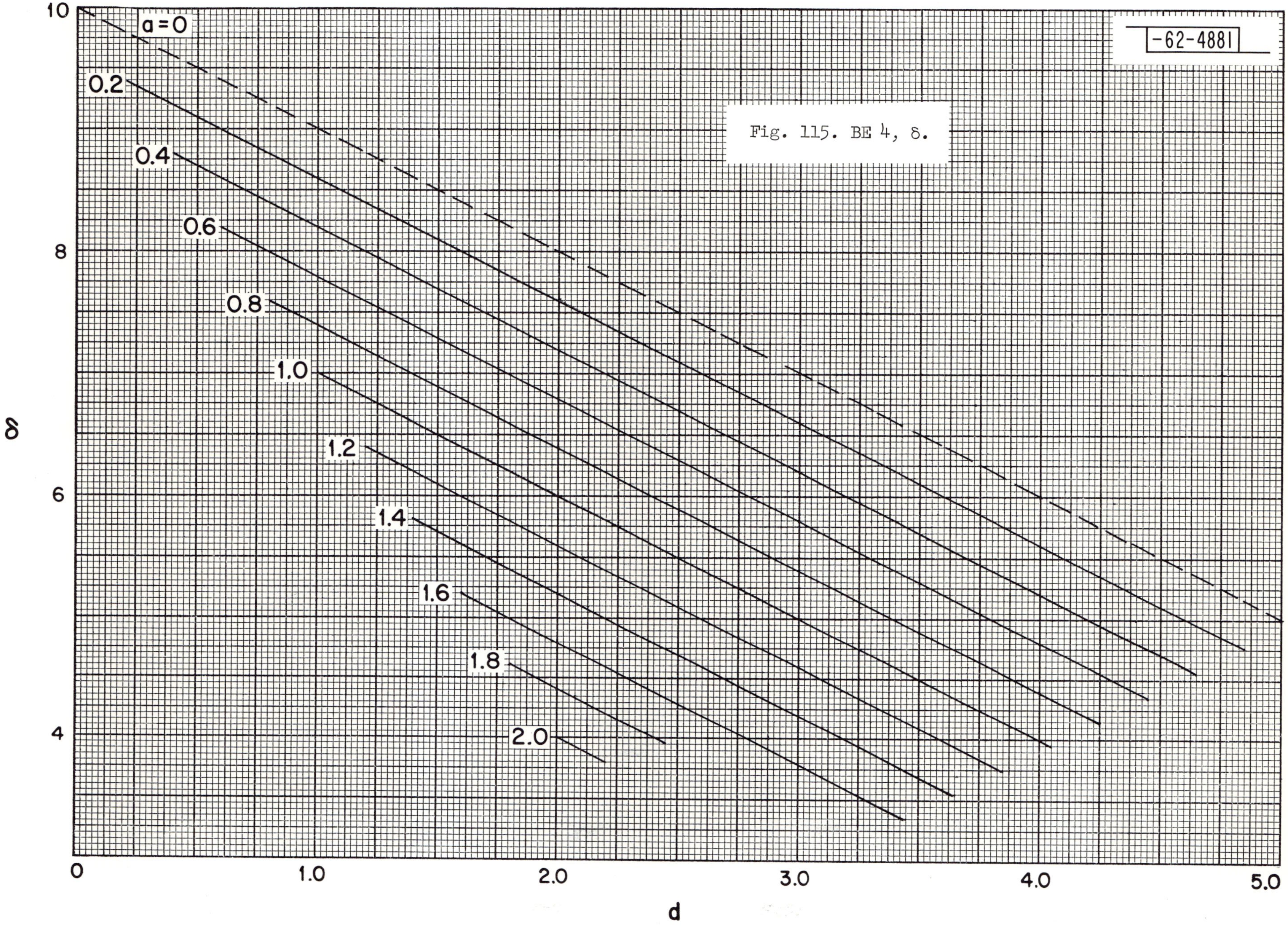


Fig. 115. BE 4, δ.

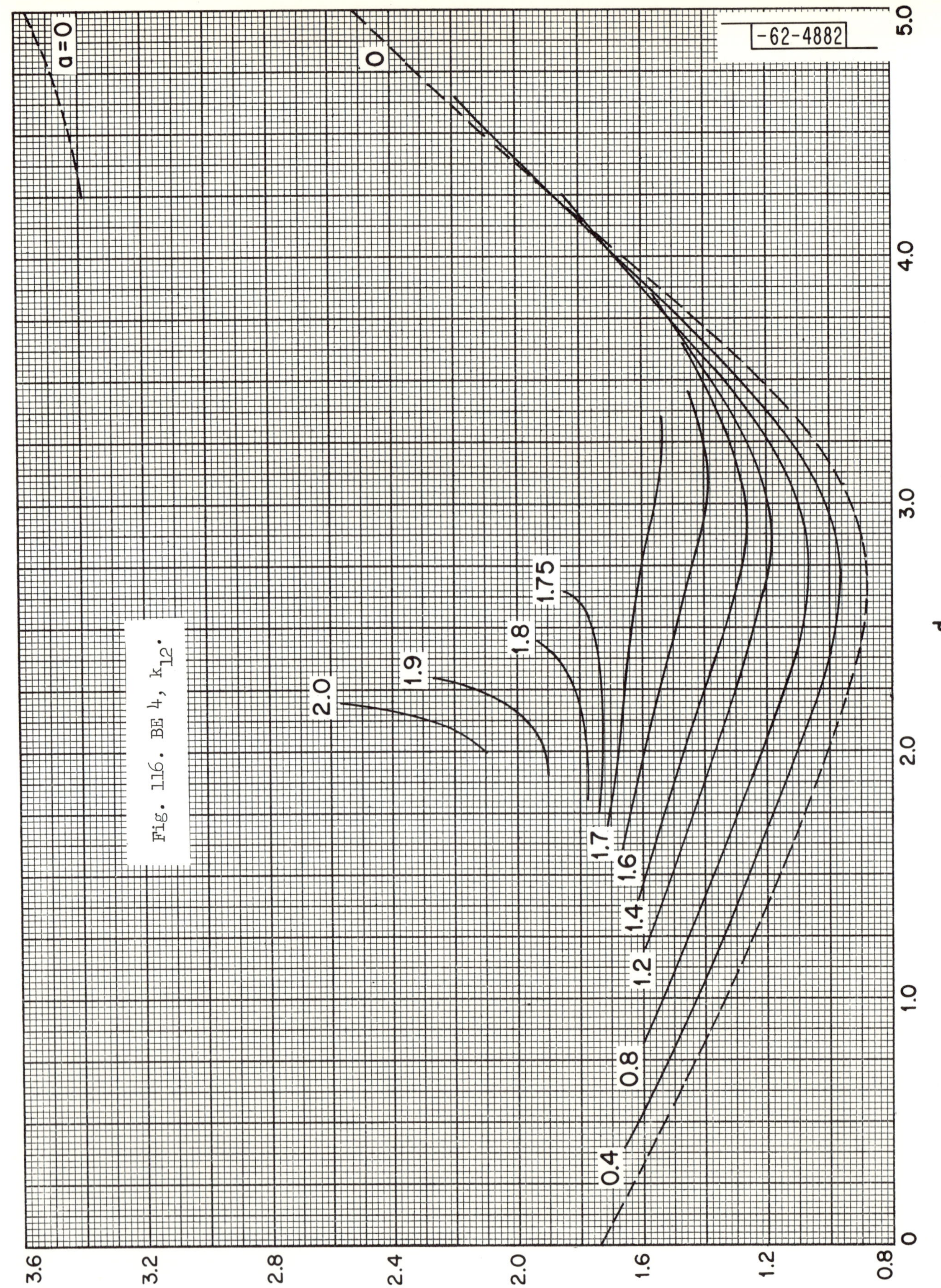


Fig. 116. BE 4, $k_{12}$.

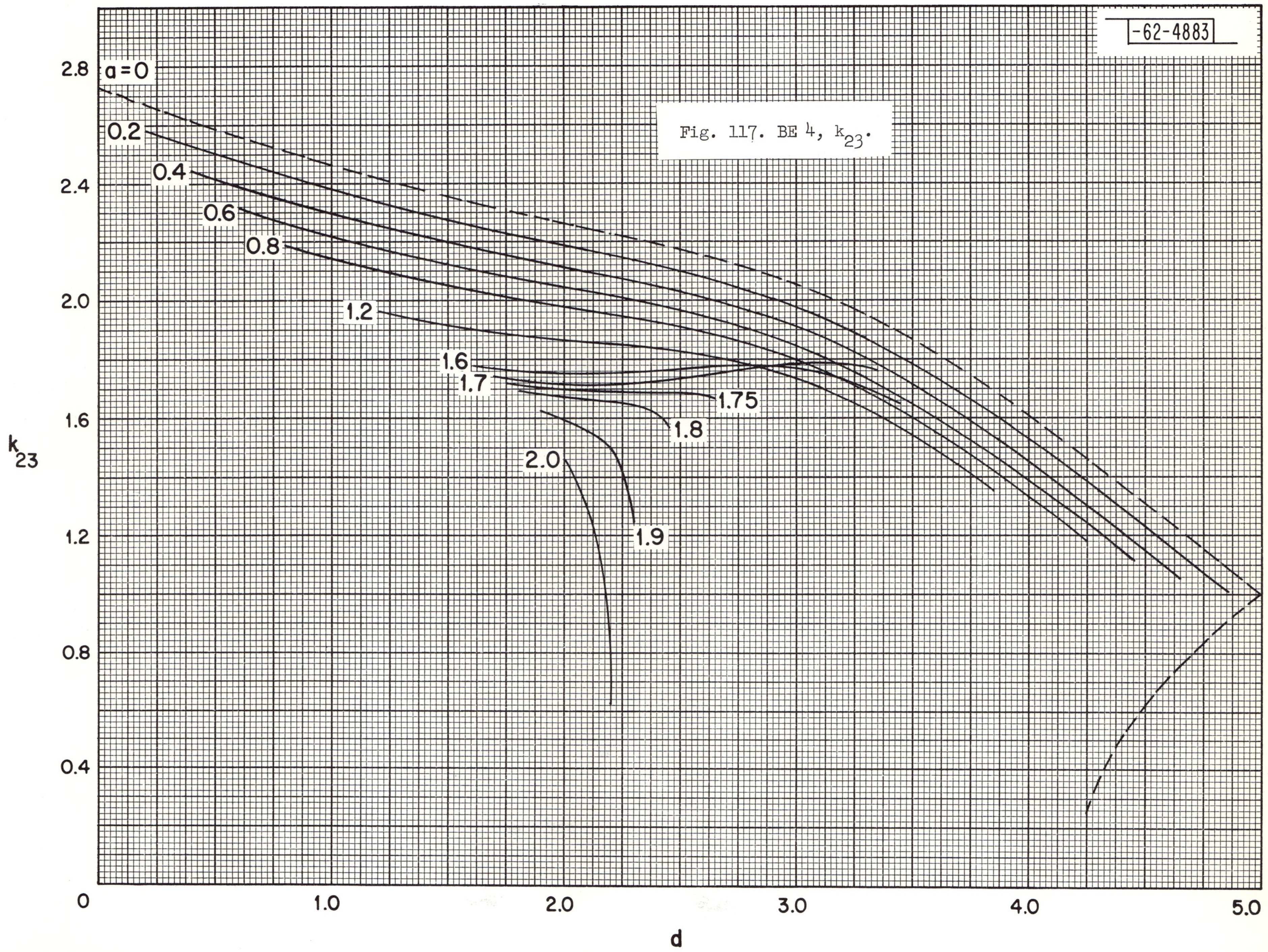


Fig. 117. BE 4, $k_{23}$.

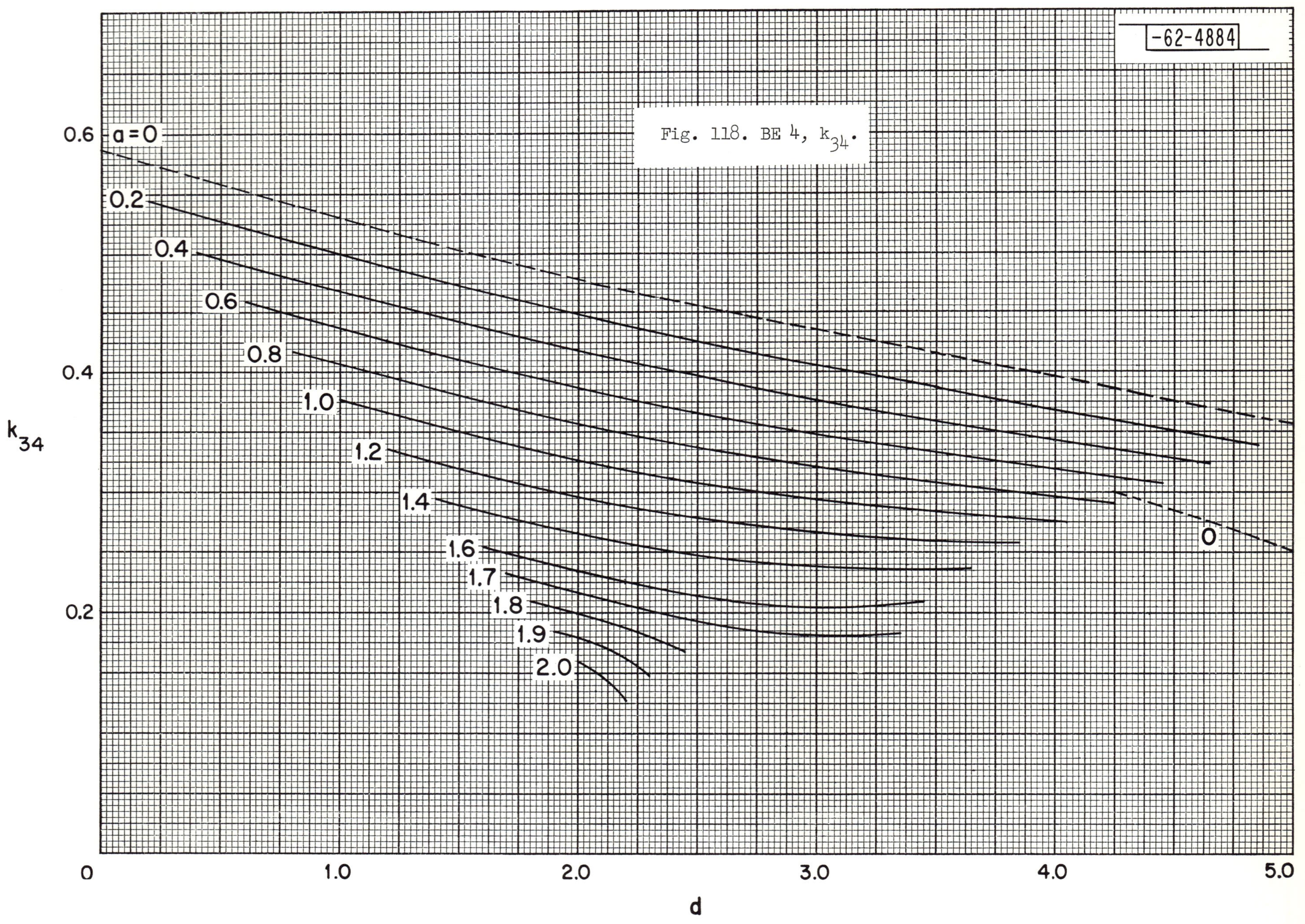


Fig. 118. BE 4, $k_{34}$.

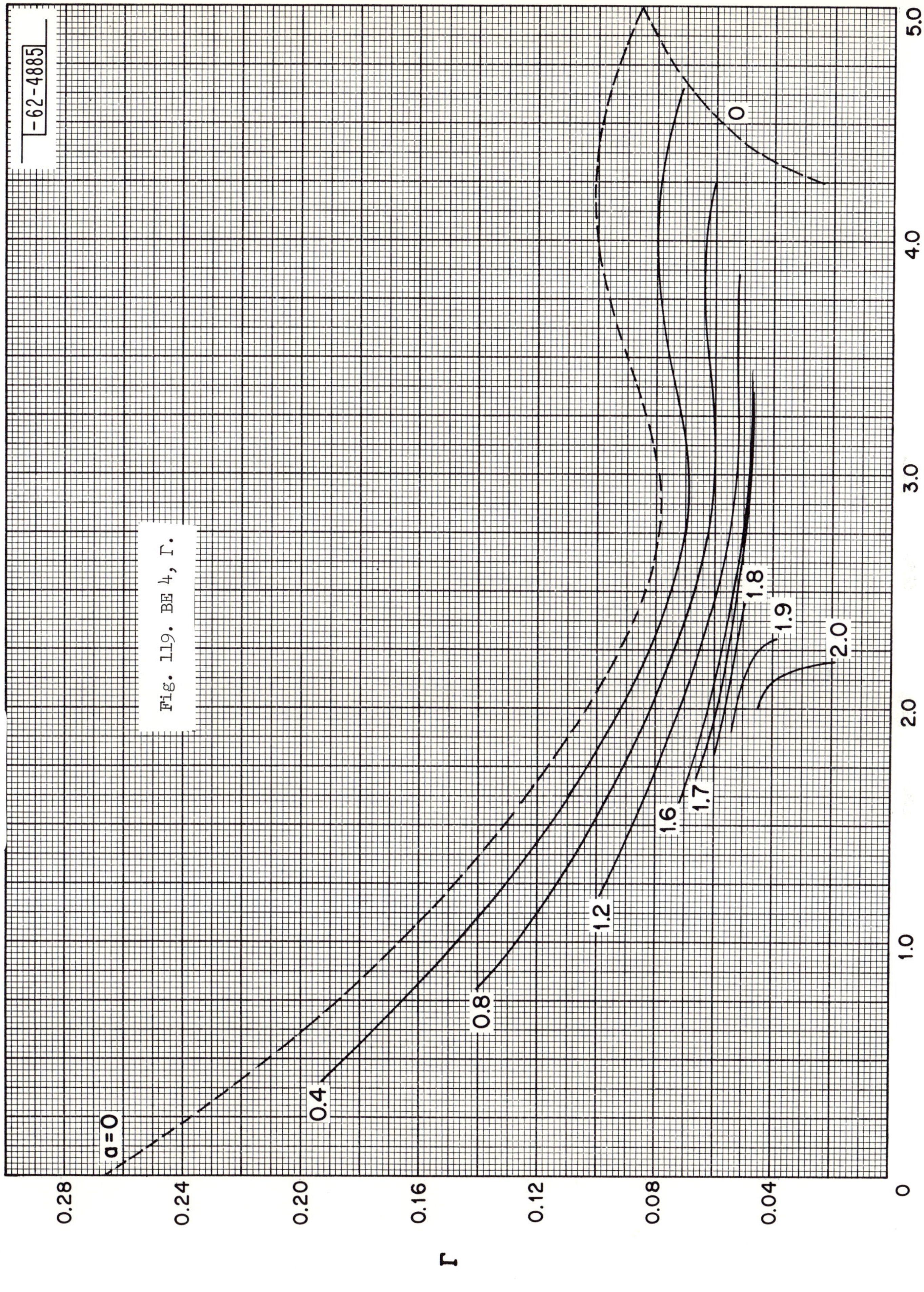
-62-4885
a=0
0.4
0.8
1.2
1.6
1.7
1.8
1.9
2.0
0
0.28
0.24
0.20
0.16
0.12
0.08
0.04
0
Γ
1.0
2.0
3.0
4.0
5.0
d

Fig. 119. BE 4, Γ.

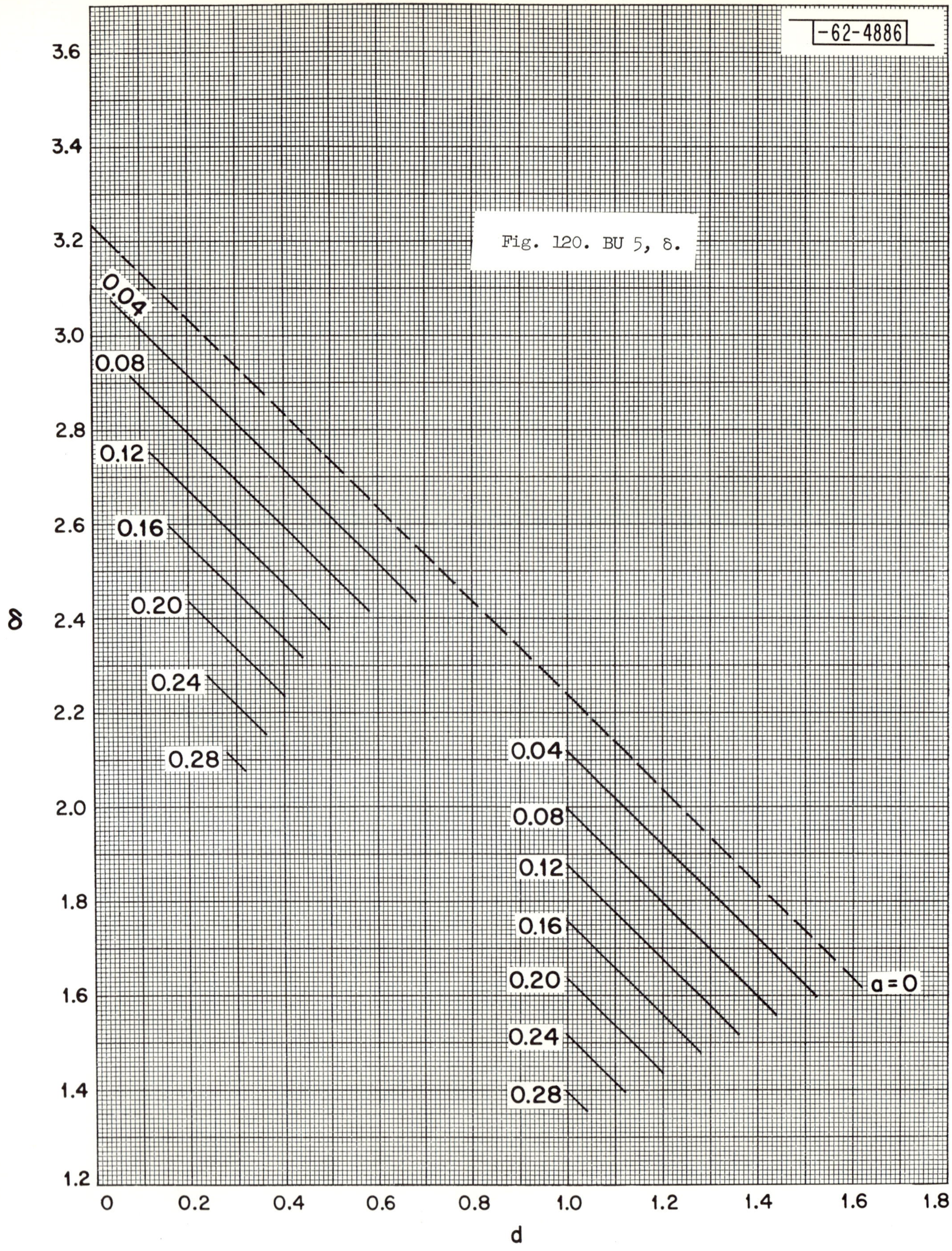
-62-4886
3.6
3.4
3.2
3.0
2.8
2.6
2.4
2.2
2.0
1.8
1.6
1.4
1.2
δ
0
0.2
0.4
0.6
0.8
1.0
1.2
1.4
1.6
1.8
d
0.04
0.08
0.12
0.16
0.20
0.24
0.28
0.04
0.08
0.12
0.16
0.20
0.24
0.28
a = 0

Fig. 120. BU 5, δ.

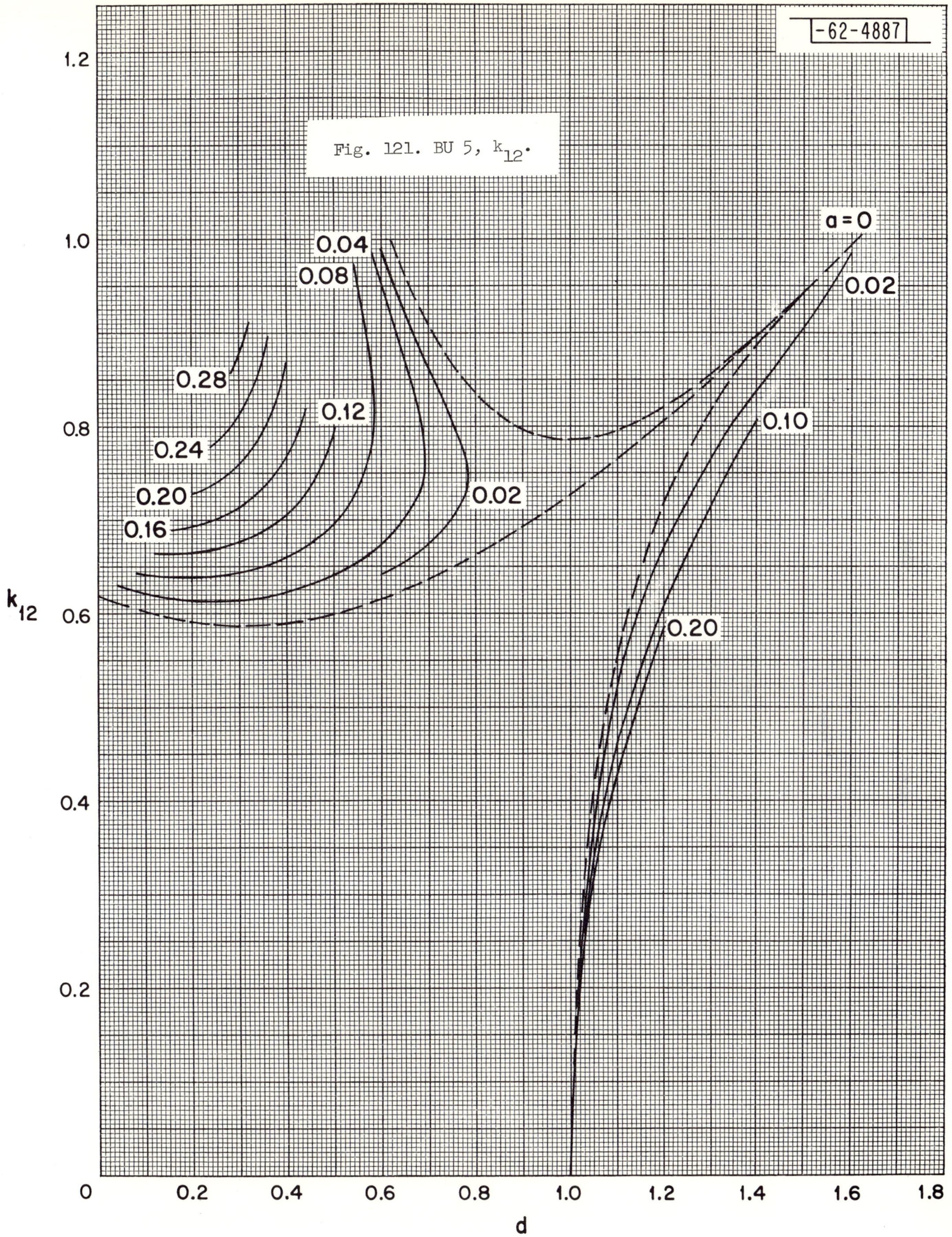
-62-4887
Fig. 121. BU 5, $k_{12}$.
a = 0
0.04
0.08
0.02
0.28
0.12
0.24
0.20
0.16
0.02
0.10
0.20
$k_{12}$
d
1.2
1.0
0.8
0.6
0.4
0.2
0
0.2
0.4
0.6
0.8
1.0
1.2
1.4
1.6
1.8

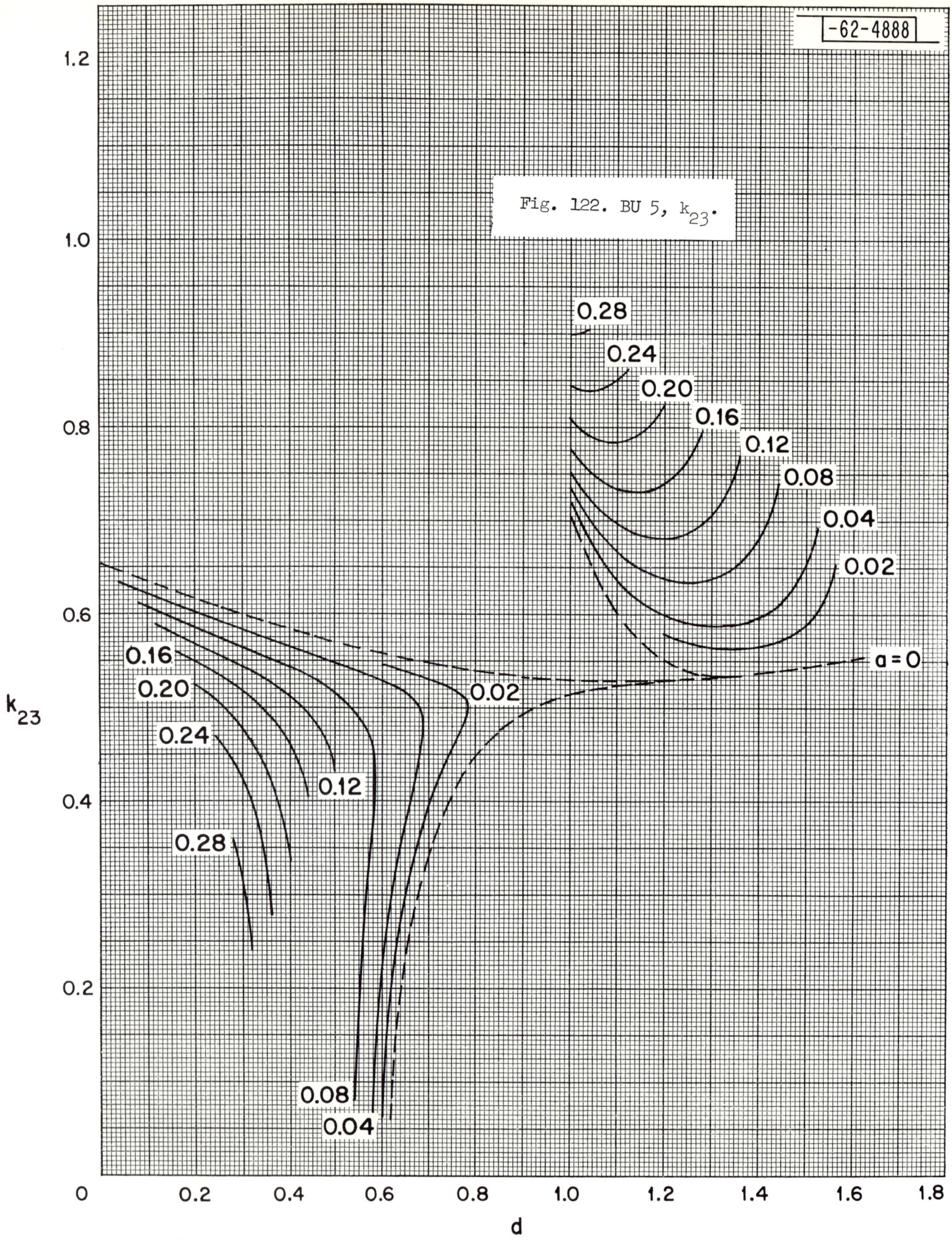


Fig. 122. BU 5, $k_{23}$.

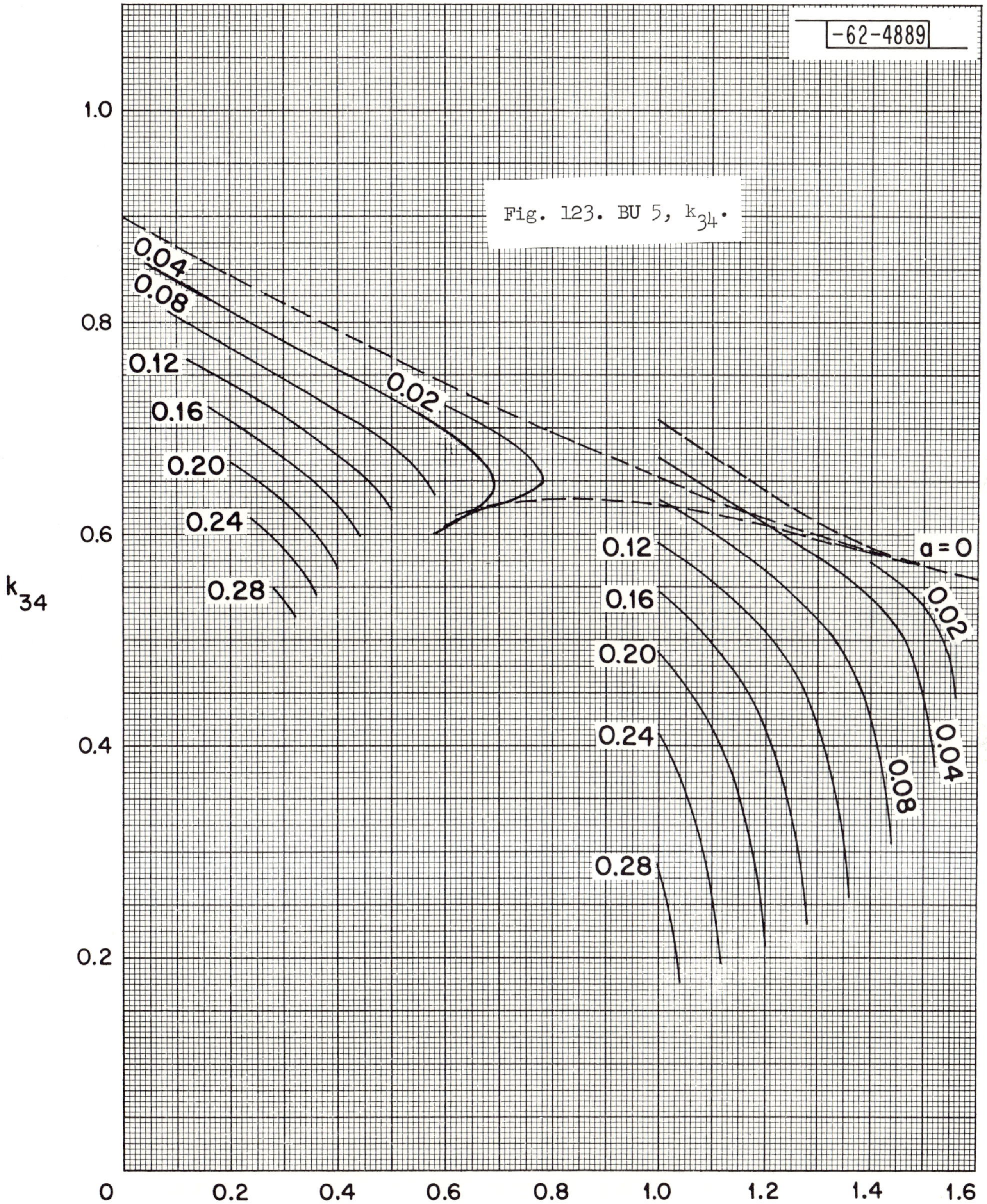
-62-4889
Fig. 123. BU 5, $k_{34}$.
1.0
0.8
0.6
0.4
0.2
$k_{34}$
0
0.2
0.4
0.6
0.8
1.0
1.2
1.4
1.6
d
0.04
0.08
0.12
0.16
0.20
0.24
0.28
0.02
0.12
0.16
0.20
0.24
0.28
a = 0
0.02
0.04
0.08

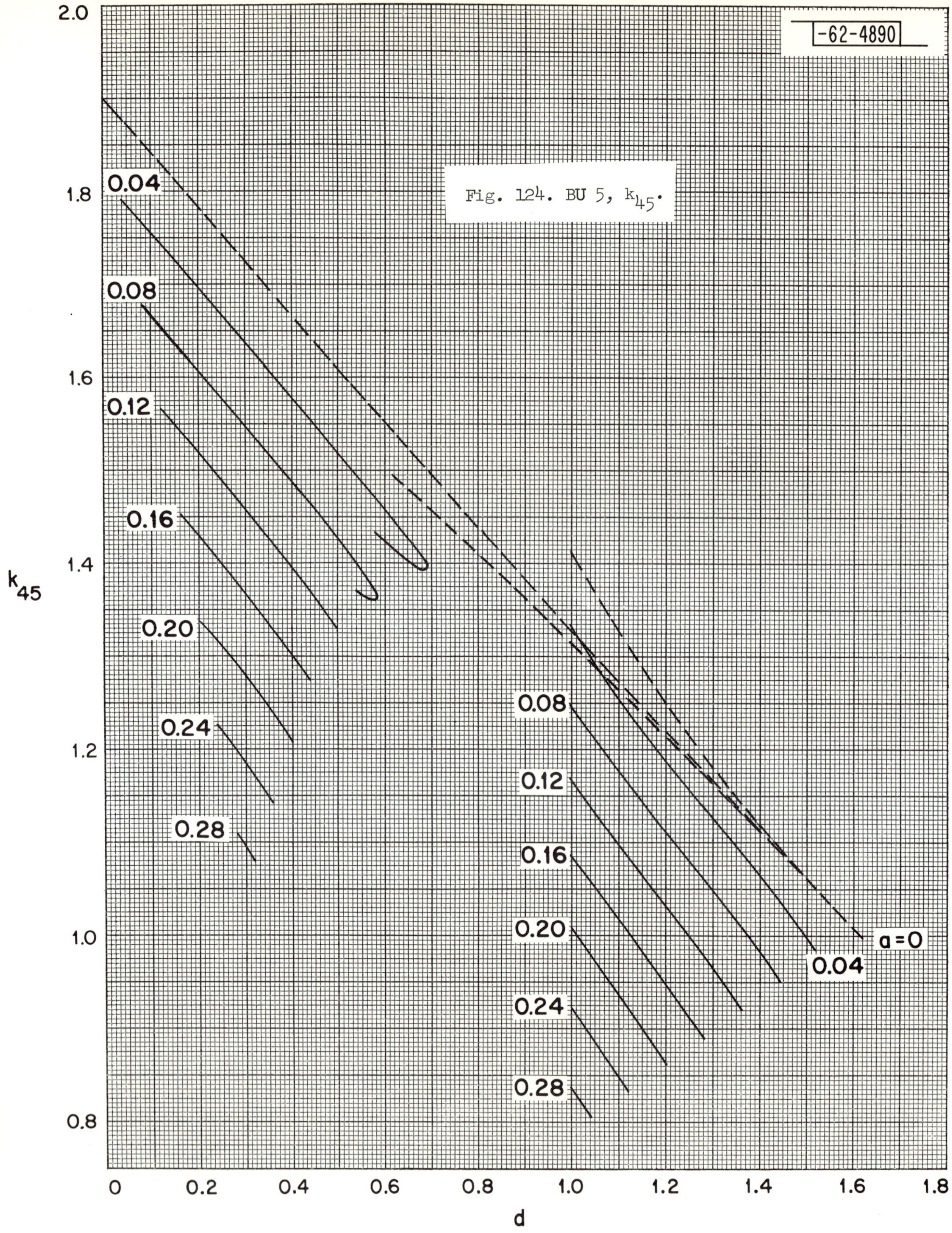

Fig. 124. BU 5, $k_{45}$.

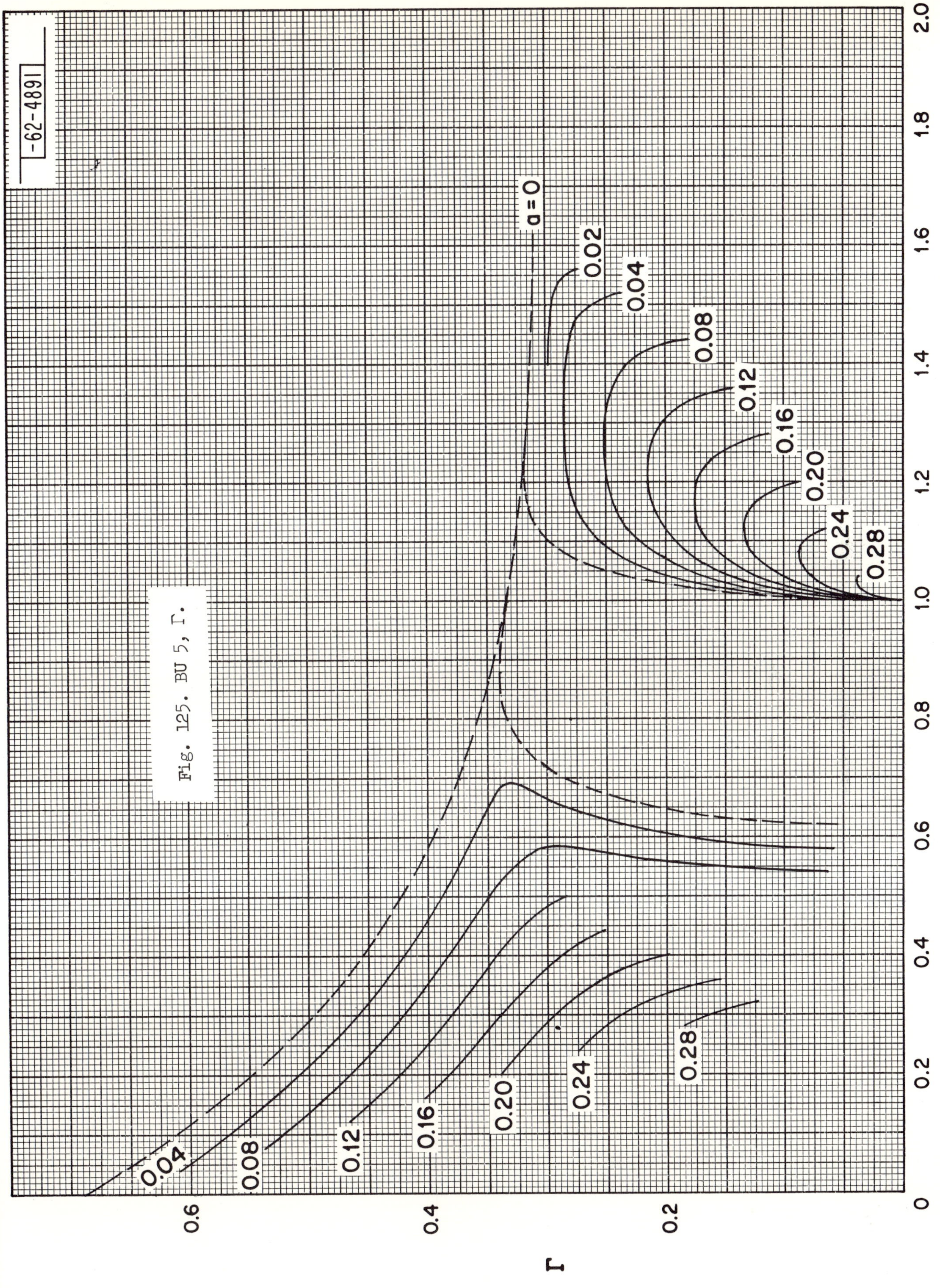
-62-4891
a=0
0.02
0.04
0.08
0.12
0.16
0.20
0.24
0.28
0.04
0.08
0.12
0.16
0.20
0.24
0.28
d
0
0.2
0.4
0.6
0.8
1.0
1.2
1.4
1.6
1.8
2.0
Γ
0.2
0.4
0.6

Fig. 125. BU 5, Γ.

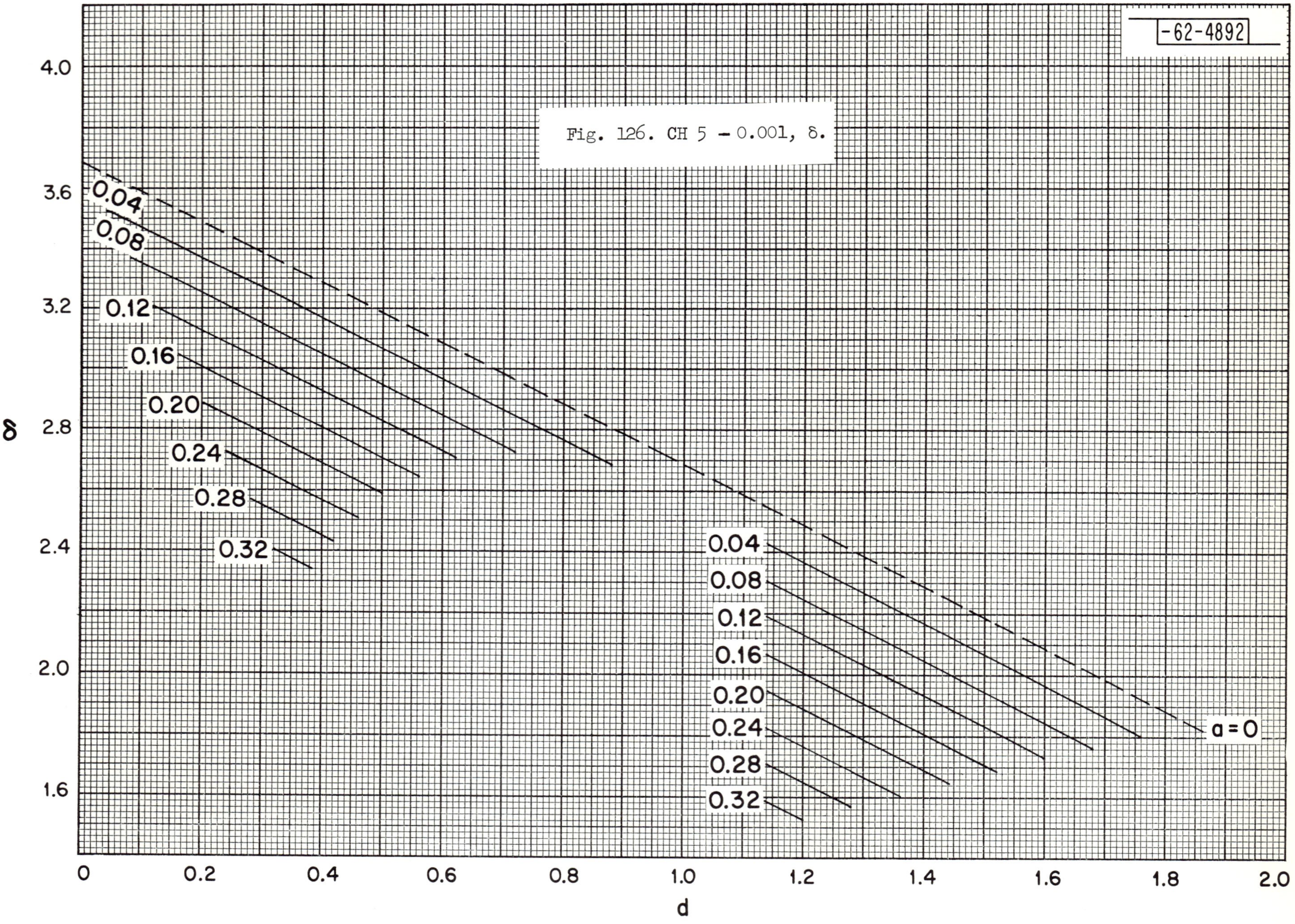
-62-4892
Fig. 126. CH 5 – 0.001, δ.
δ
4.0
3.6
3.2
2.8
2.4
2.0
1.6
0.04
0.08
0.12
0.16
0.20
0.24
0.28
0.32
a = 0
0
0.2
0.4
0.6
0.8
1.0
1.2
1.4
1.6
1.8
2.0
d

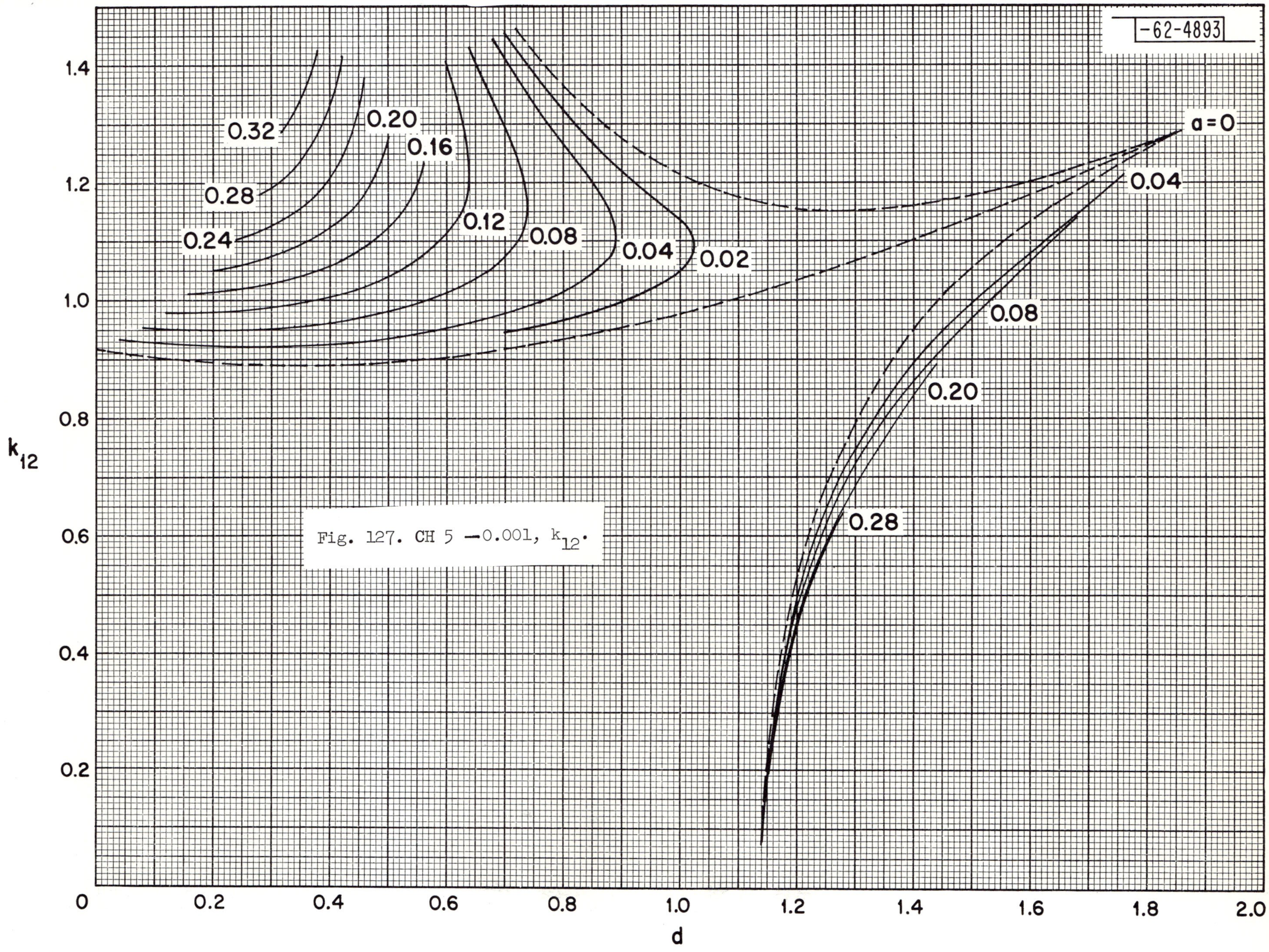


Fig. 127. CH 5 —0.001, $k_{12}$.

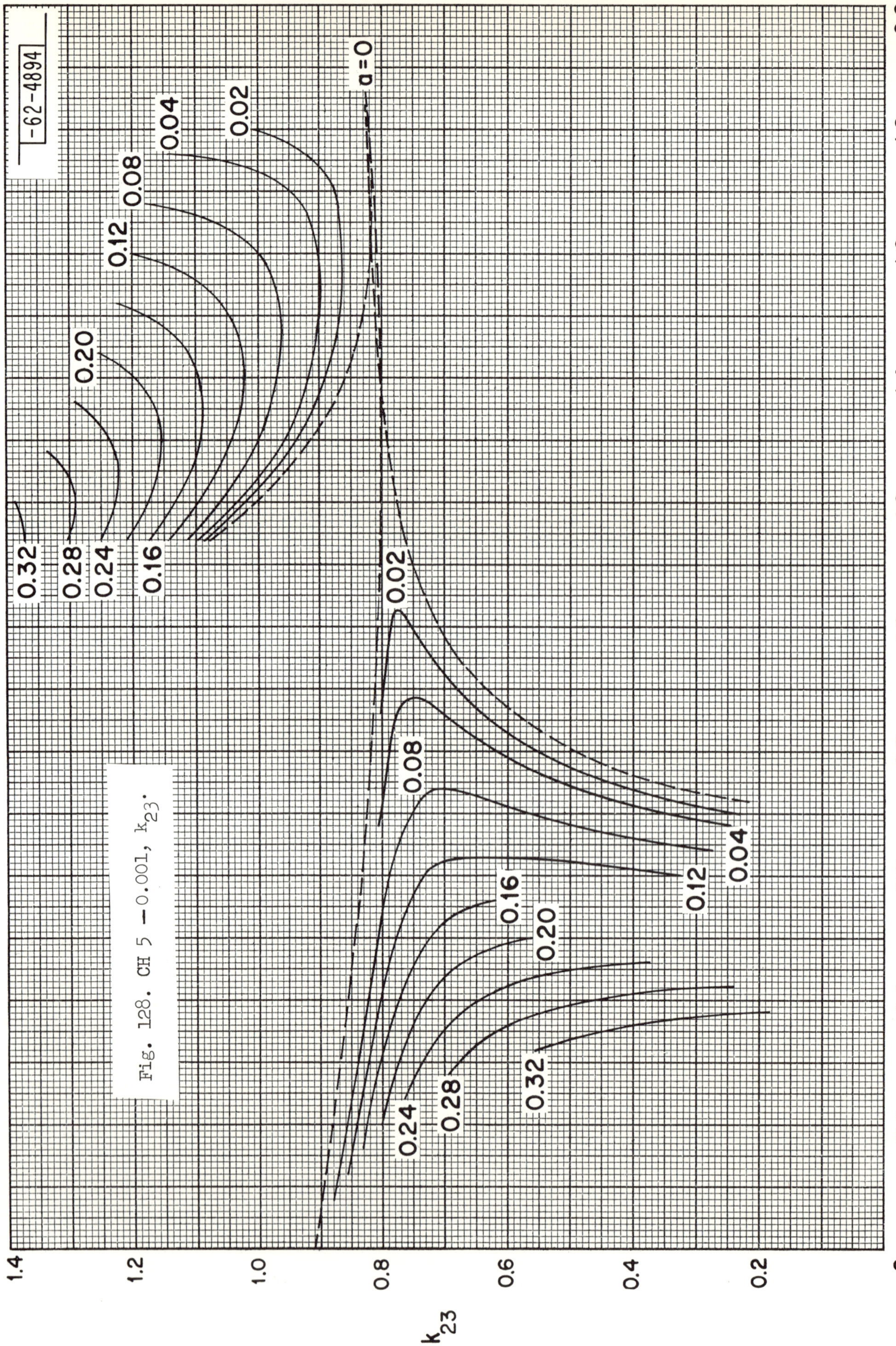


Fig. 128. CH 5 — 0.001, $k_{23}$.

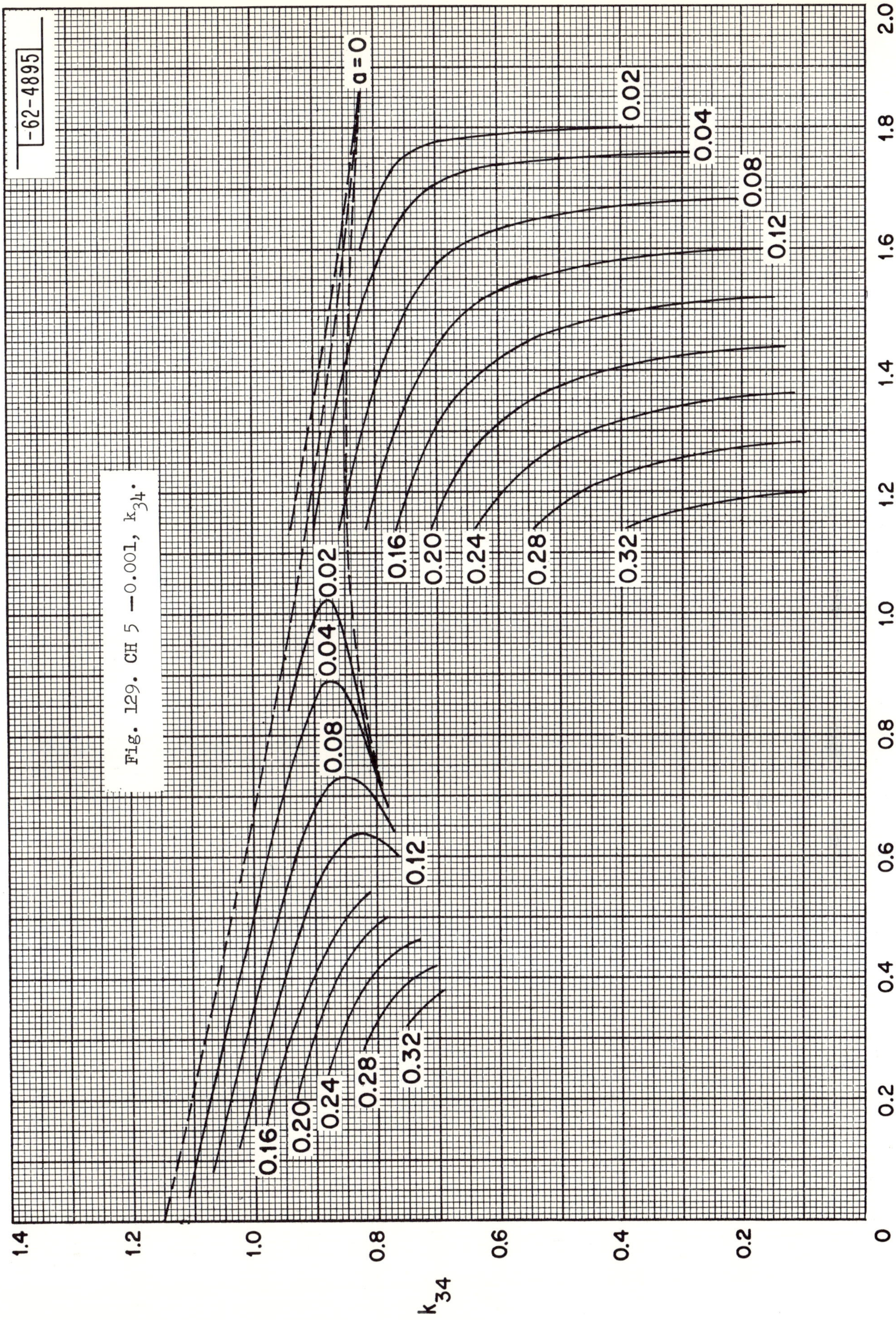


Fig. 129. CH 5 —0.001, $k_{34}$.

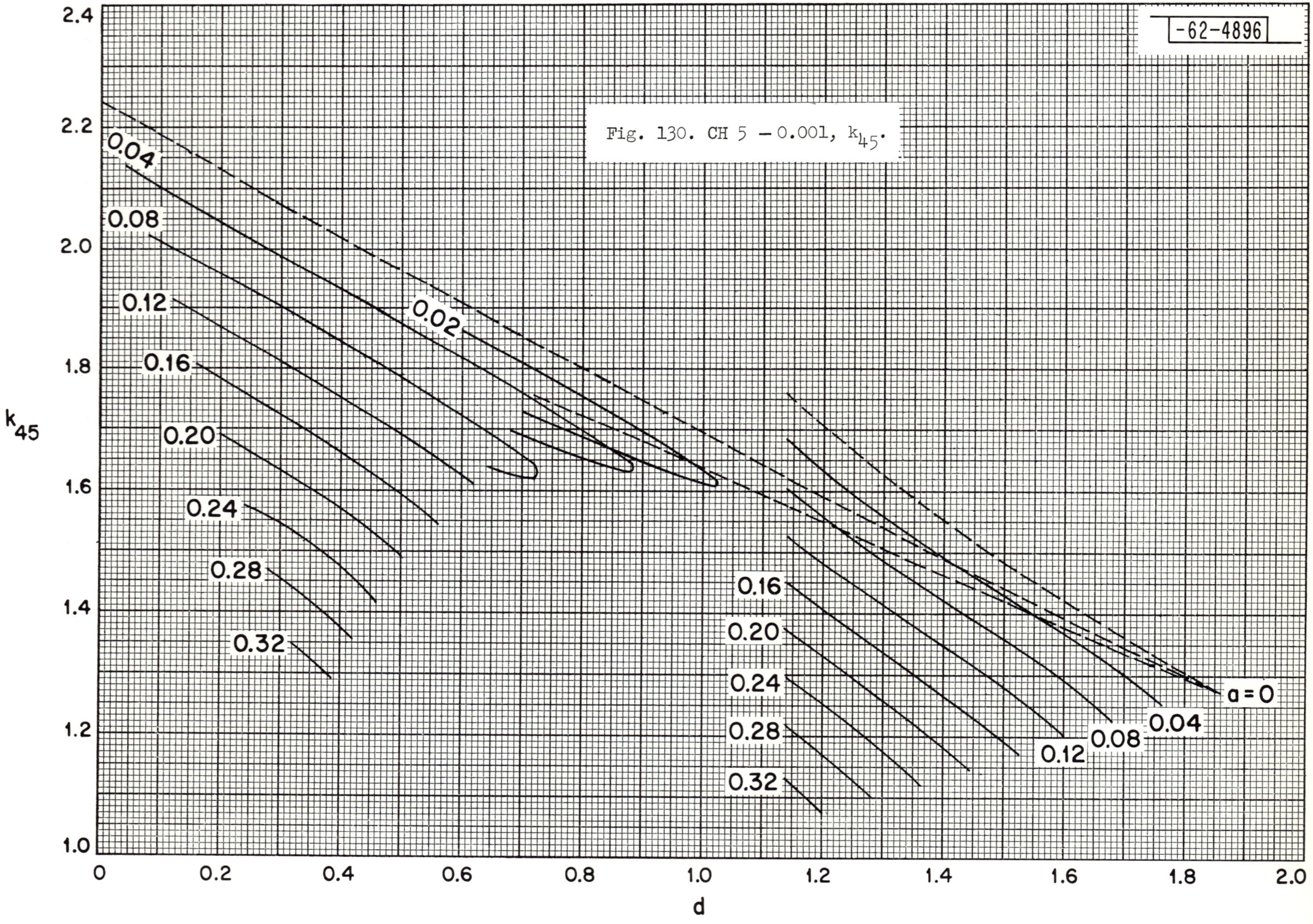


Fig. 130. CH 5 — 0.001, $k_{45}$.

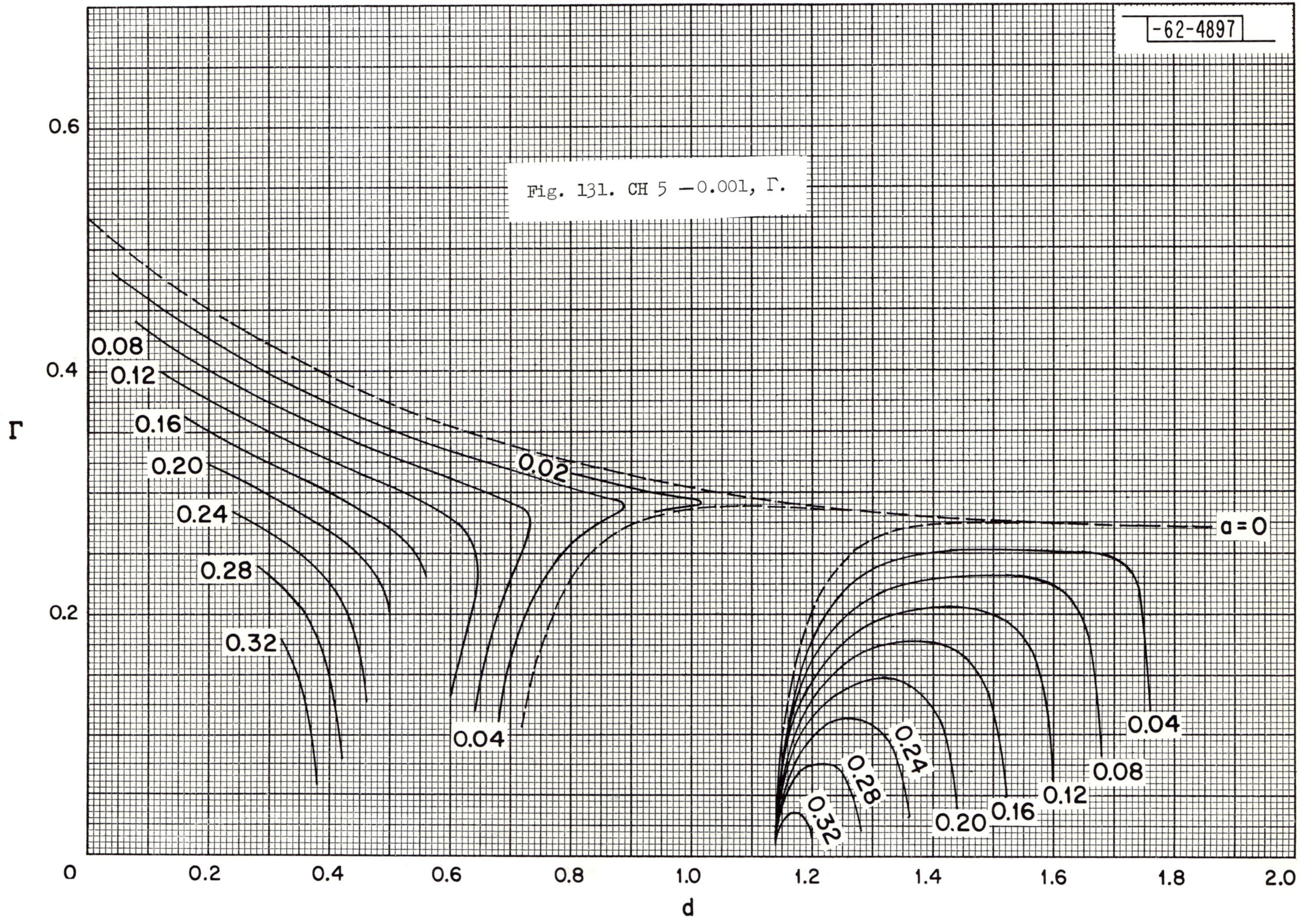


Fig. 131. CH 5 —0.001, Γ.

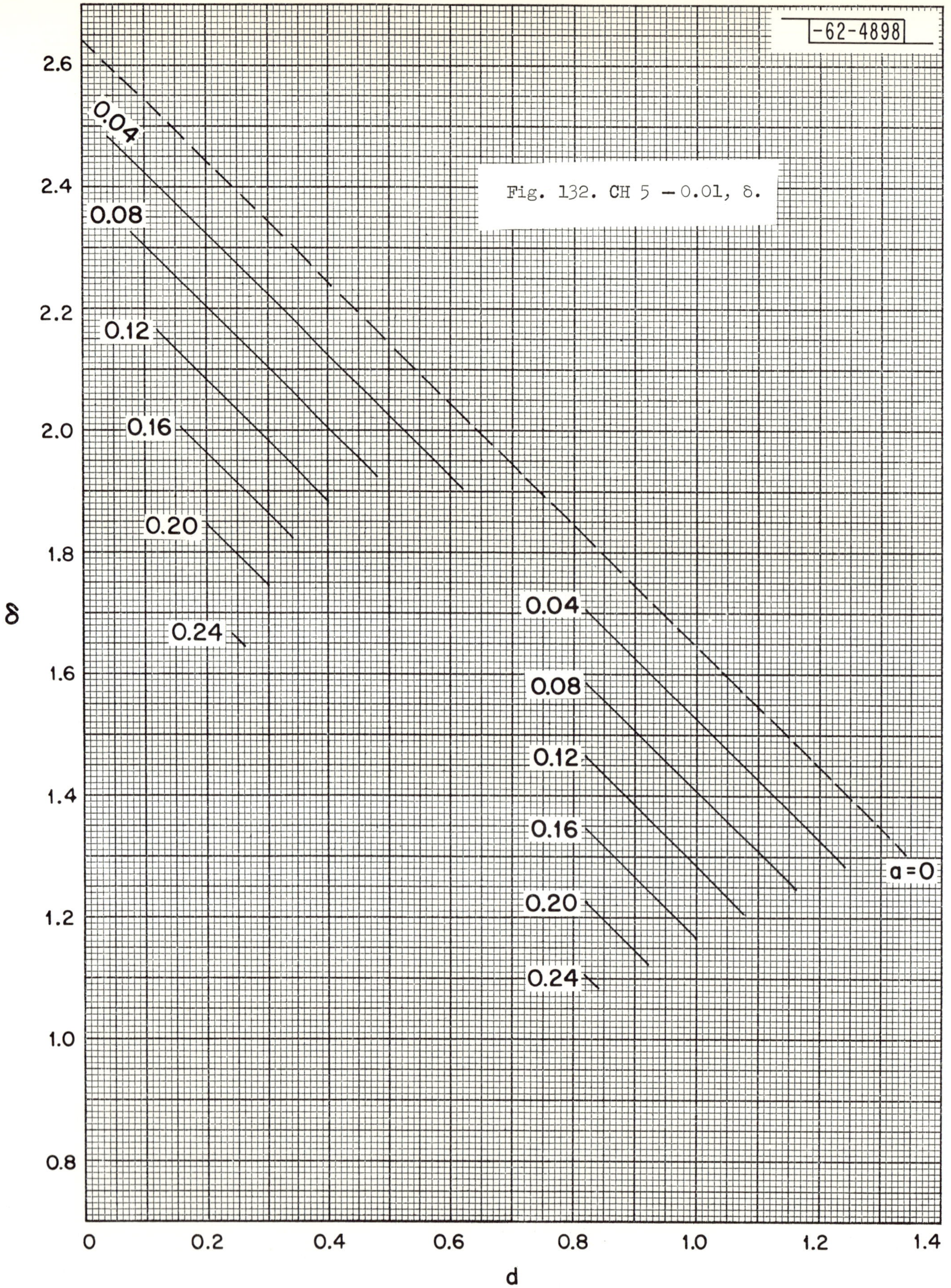

-62-4898
Fig. 132. CH 5 — 0.01, δ.
2.6
2.4
2.2
2.0
1.8
1.6
1.4
1.2
1.0
0.8
δ
0.04
0.08
0.12
0.16
0.20
0.24
0.04
0.08
0.12
0.16
0.20
0.24
a = 0
0
0.2
0.4
0.6
0.8
1.0
1.2
1.4
d

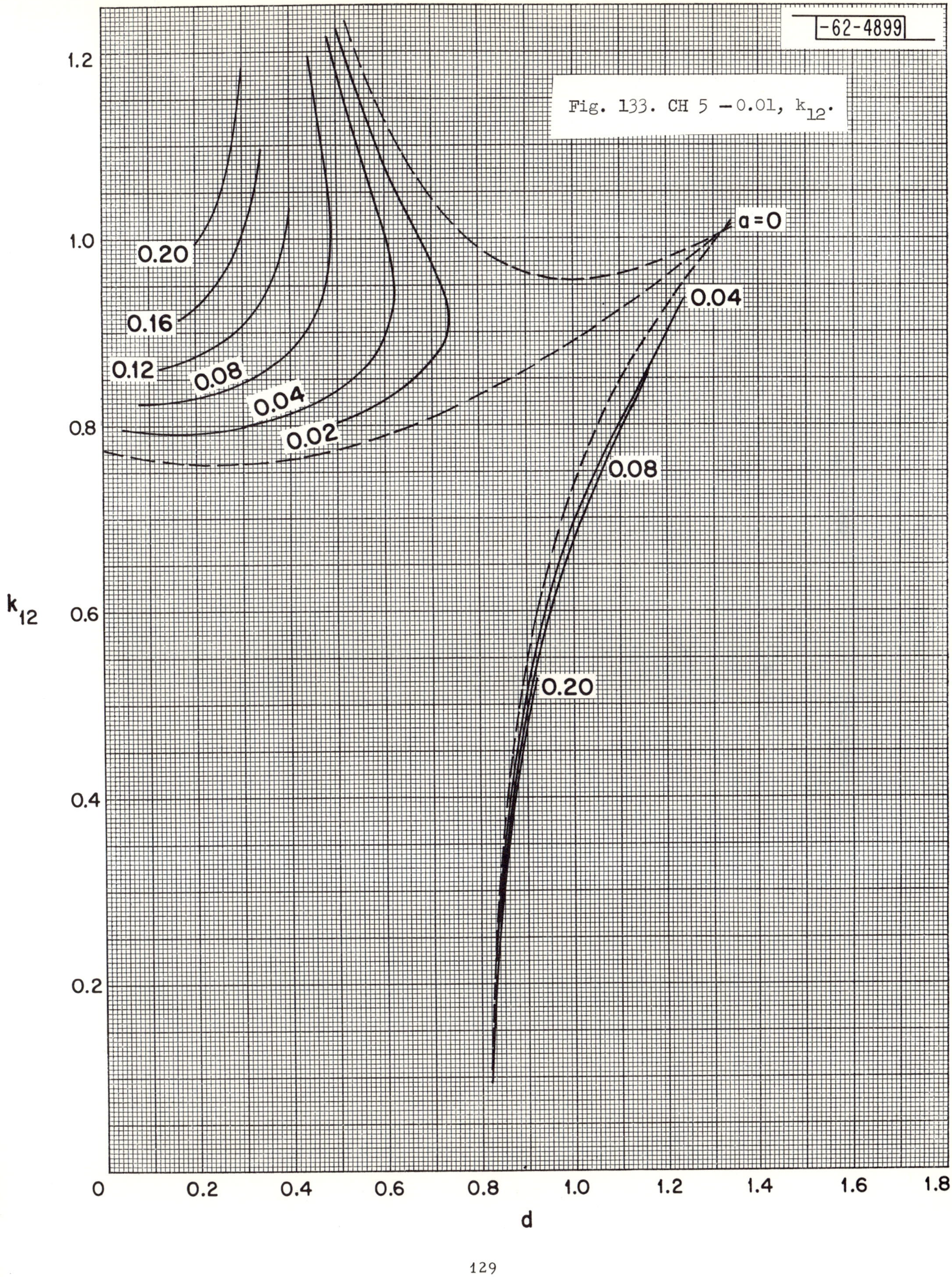
-62-4899
Fig. 133. CH 5 —0.01, $k_{12}$.
a=0
0.20
0.16
0.12
0.08
0.04
0.02
0.04
0.08
0.20
1.2
1.0
0.8
0.6
0.4
0.2
$k_{12}$
0
0.2
0.4
0.6
0.8
1.0
1.2
1.4
1.6
1.8
d

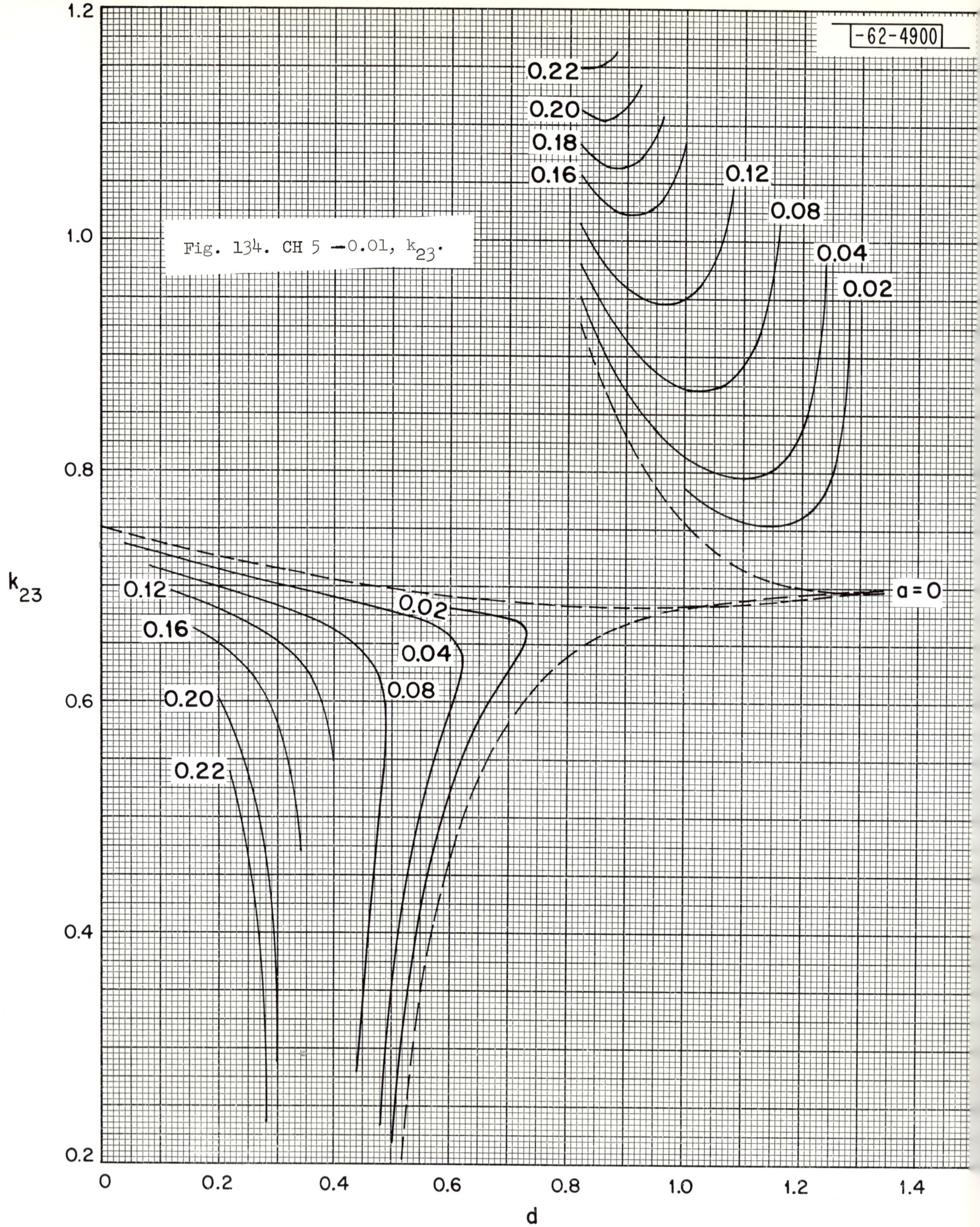

Fig. 134. CH 5 —0.01, $k_{23}$.

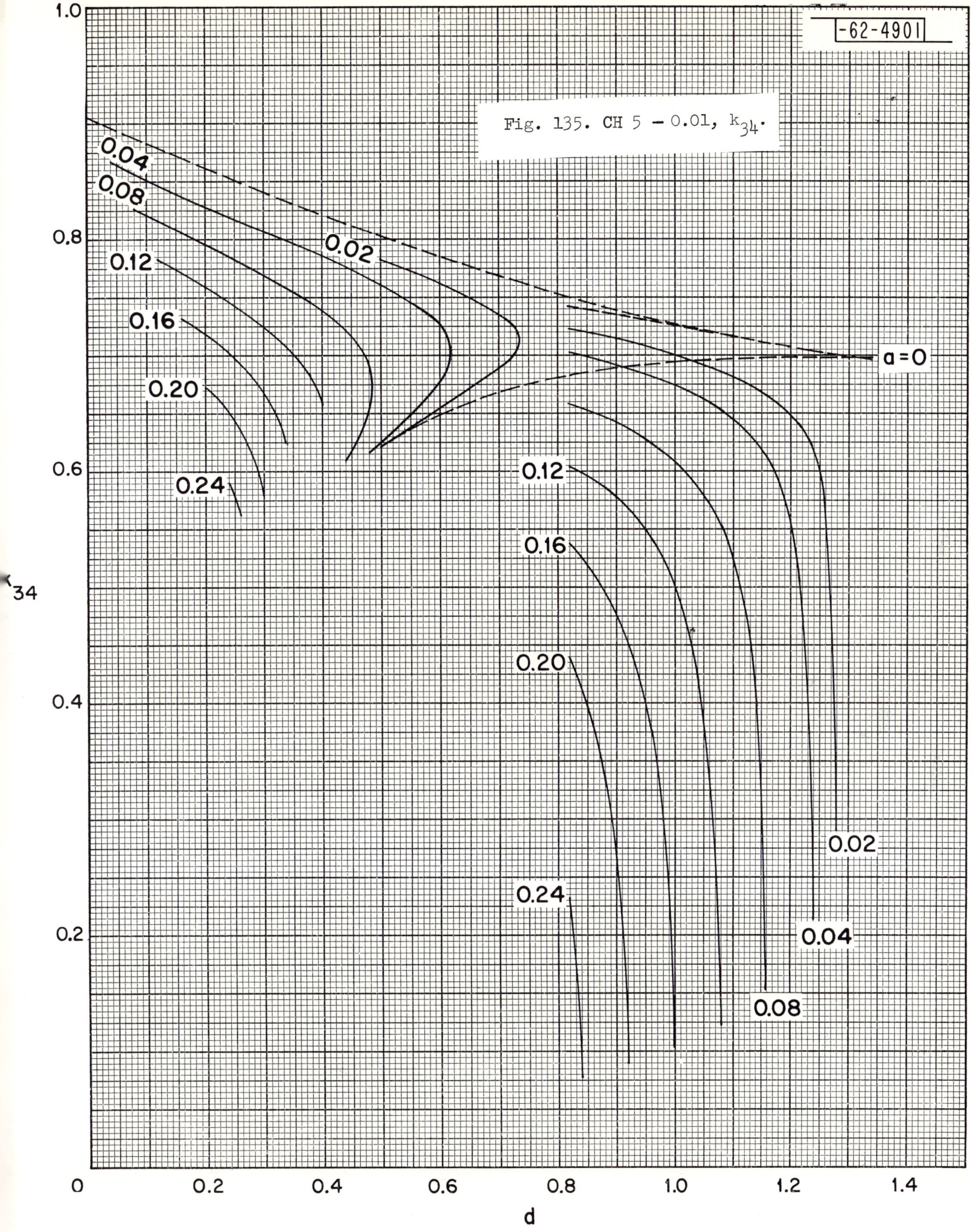


Fig. 135. CH 5 — 0.01, $k_{34}$.

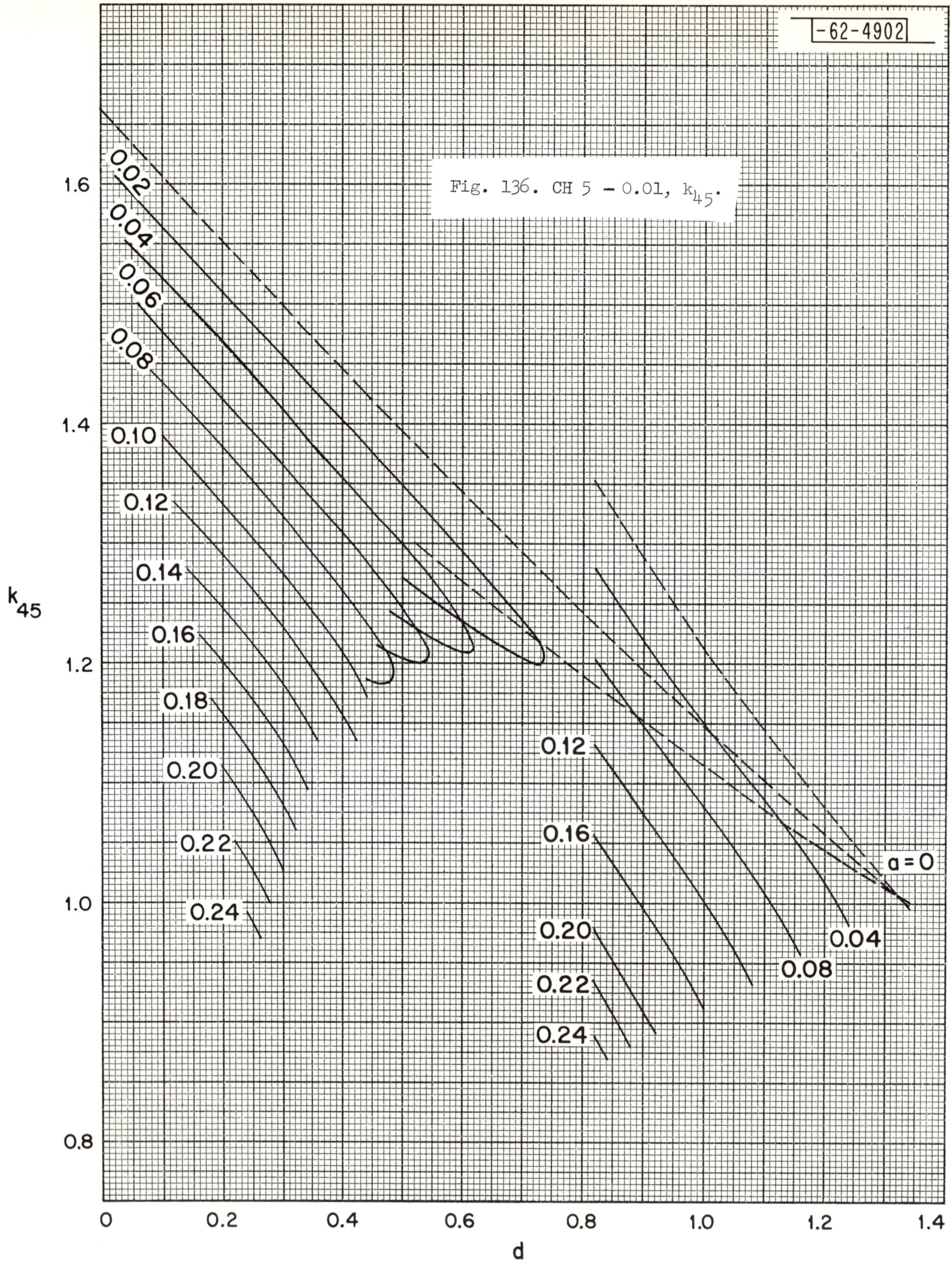


Fig. 136. CH 5 – 0.01, $k_{45}$.

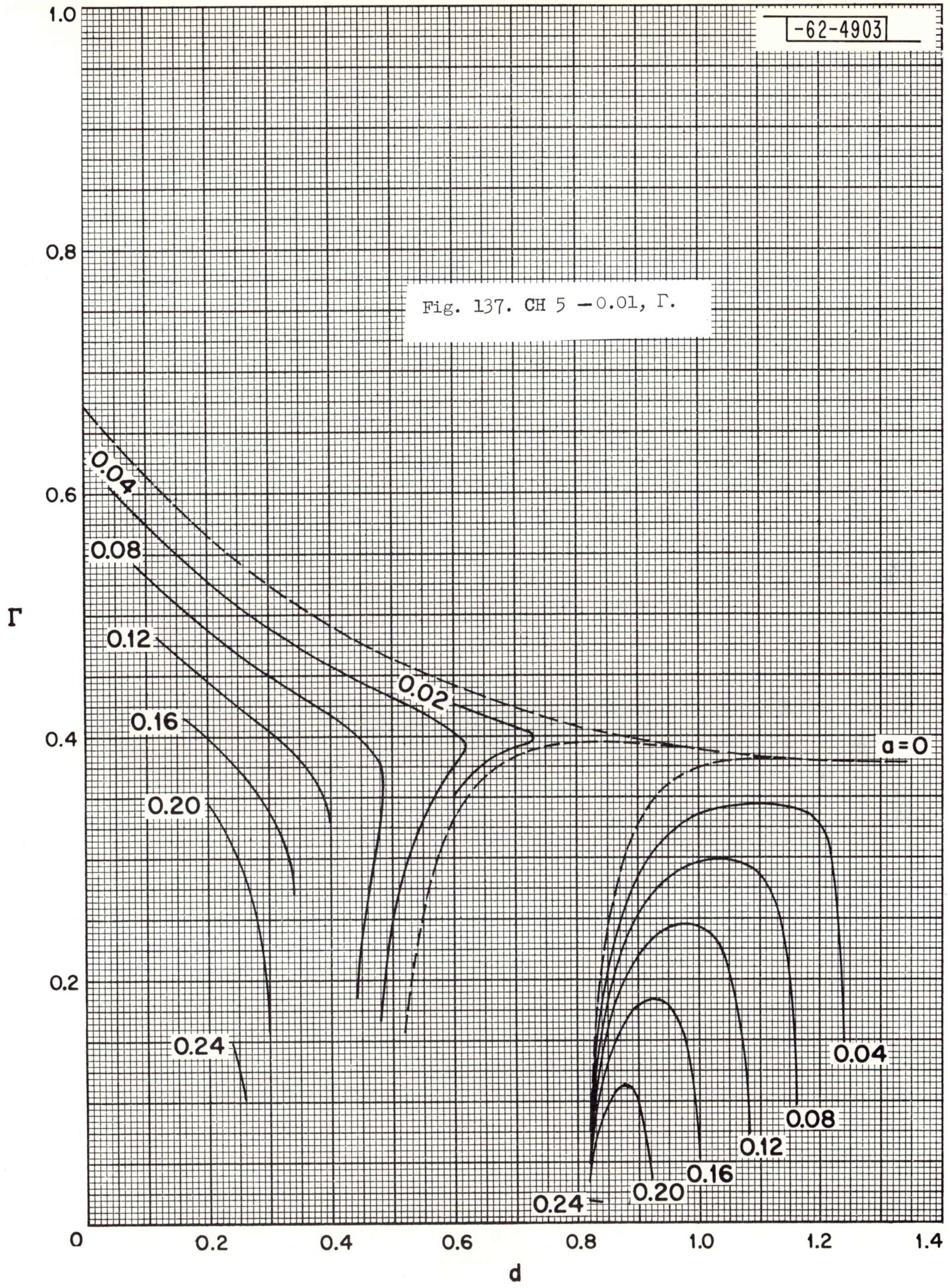


Fig. 137. CH 5 —0.01, Γ.

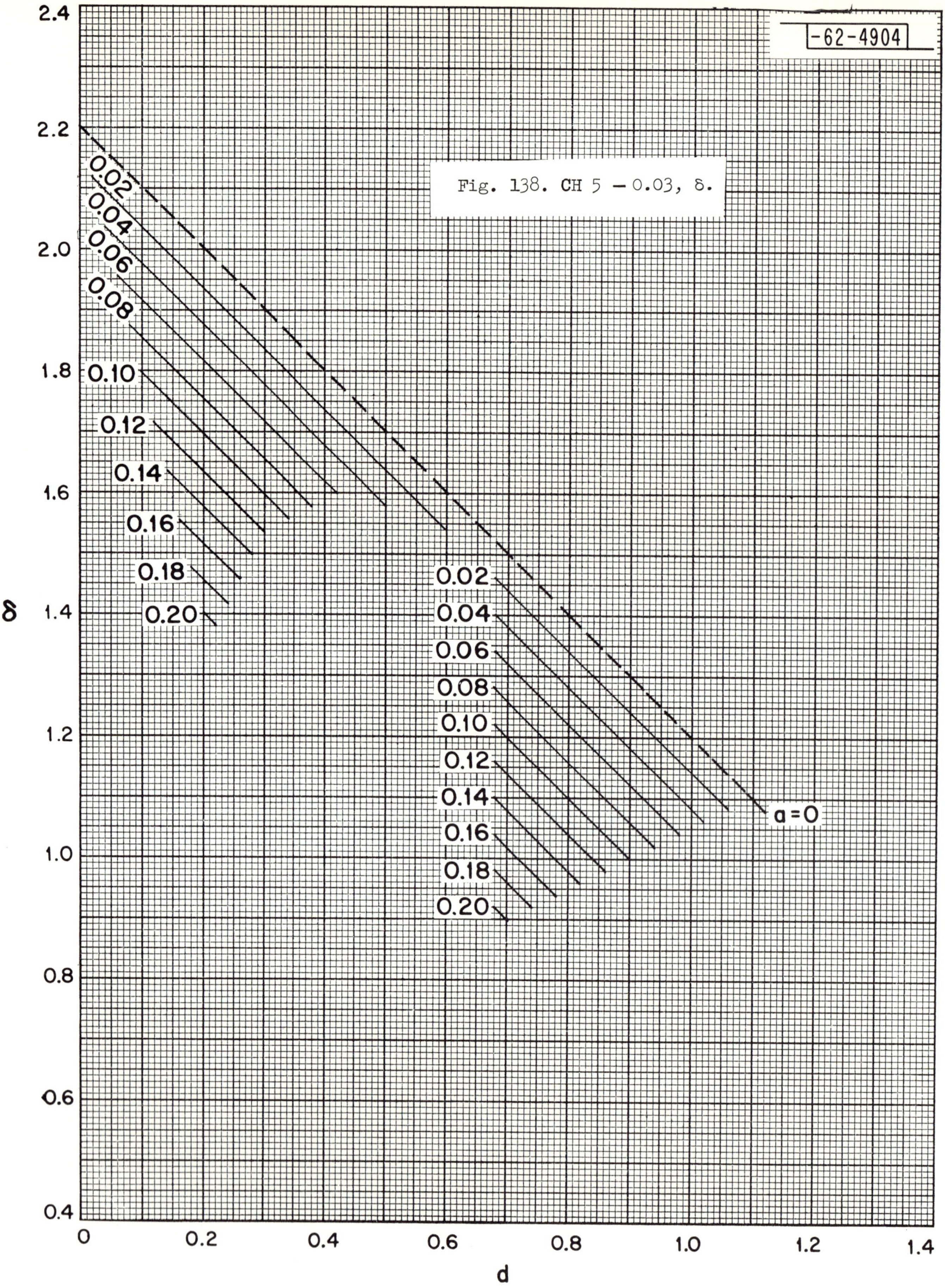


Fig. 138. CH 5 — 0.03, δ.

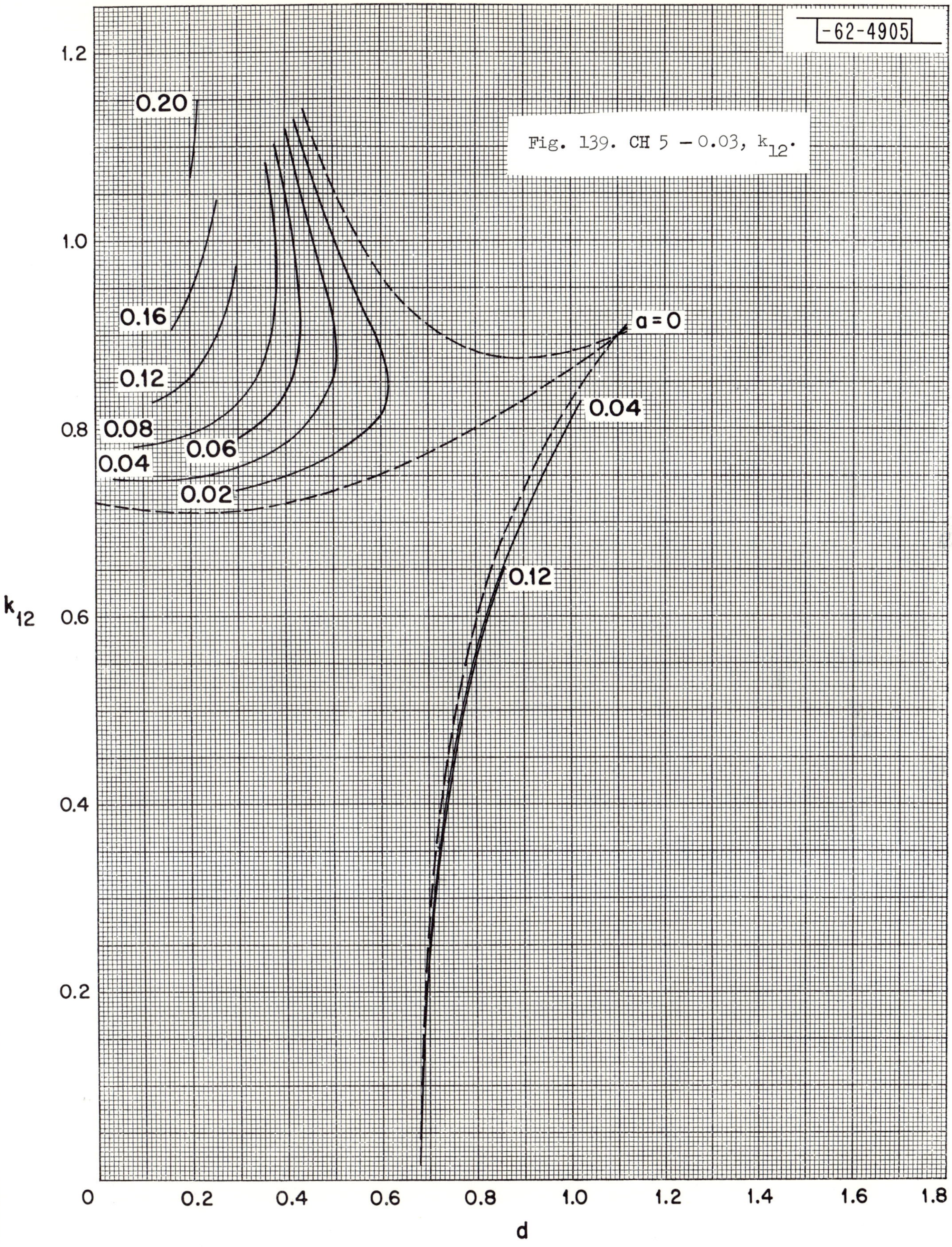


Fig. 139. CH 5 – 0.03, $k_{12}$.

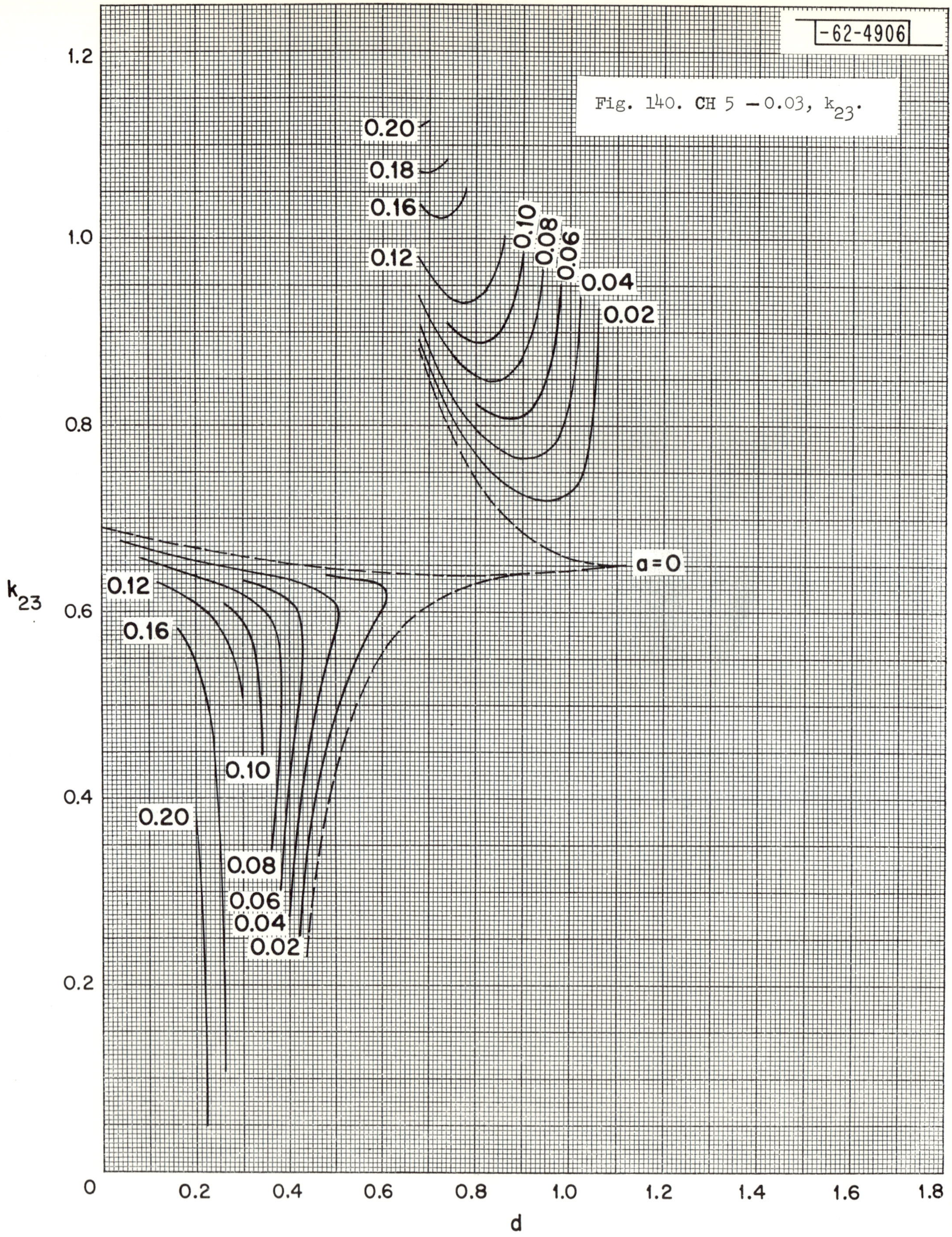


Fig. 140. CH 5 — 0.03, $k_{23}$.

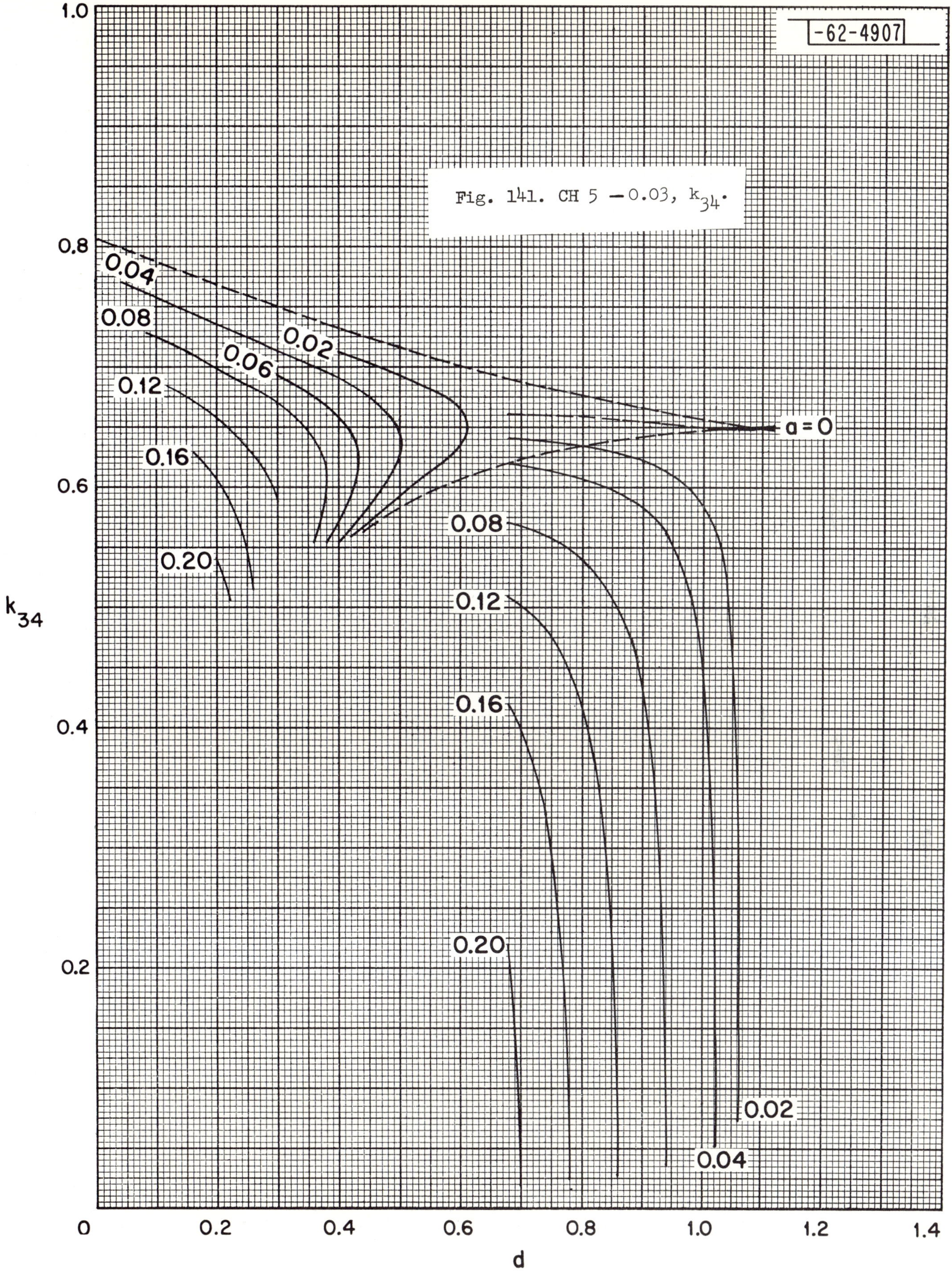


Fig. 141. CH 5 — 0.03, $k_{34}$.

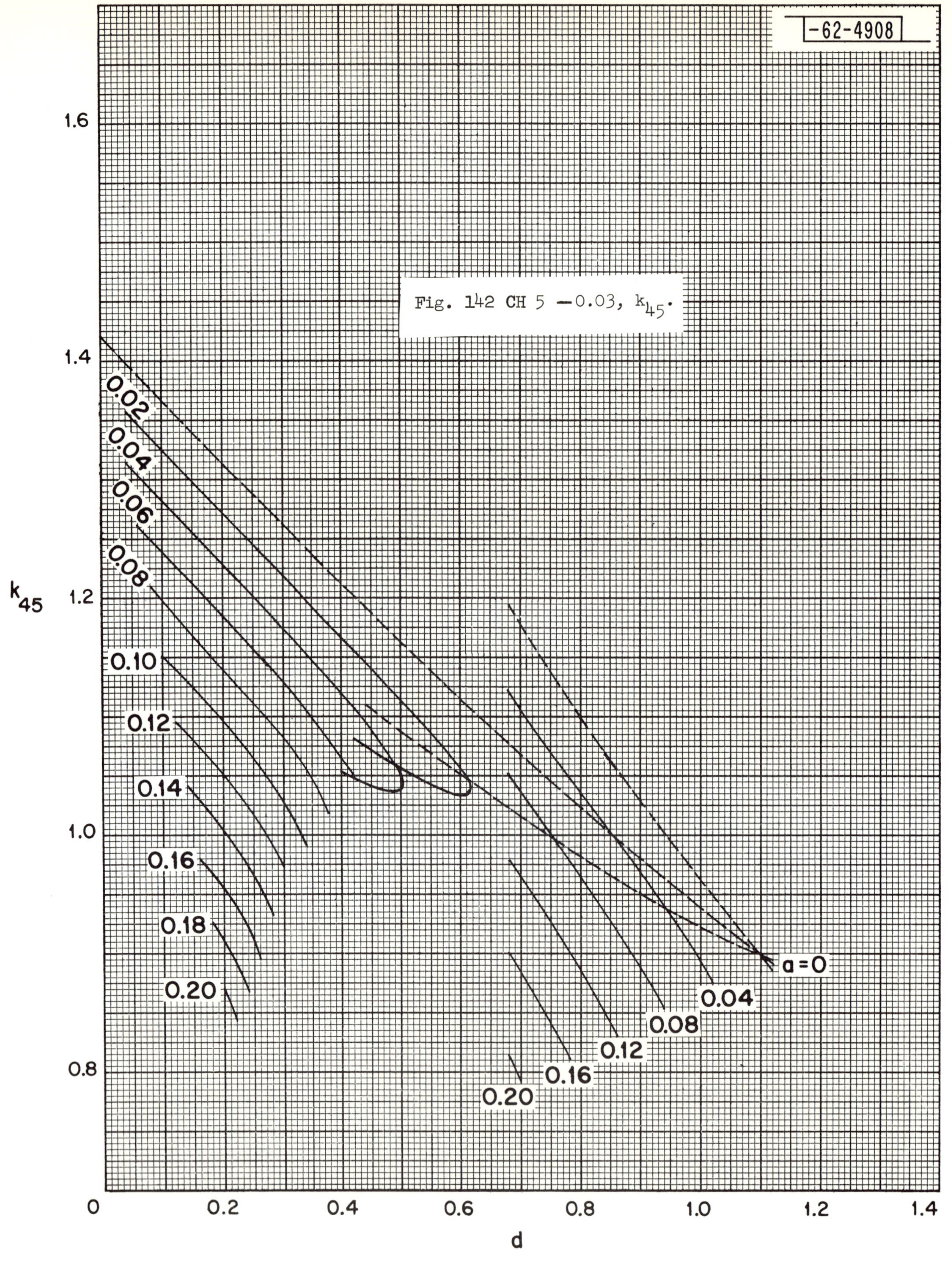
-62-4908
Fig. 142 CH 5 —0.03, $k_{45}$.
$k_{45}$
1.6
1.4
1.2
1.0
0.8
0.02
0.04
0.06
0.08
0.10
0.12
0.14
0.16
0.18
0.20
a=0
0.04
0.08
0.12
0.16
0.20
0
0.2
0.4
0.6
0.8
1.0
1.2
1.4
d

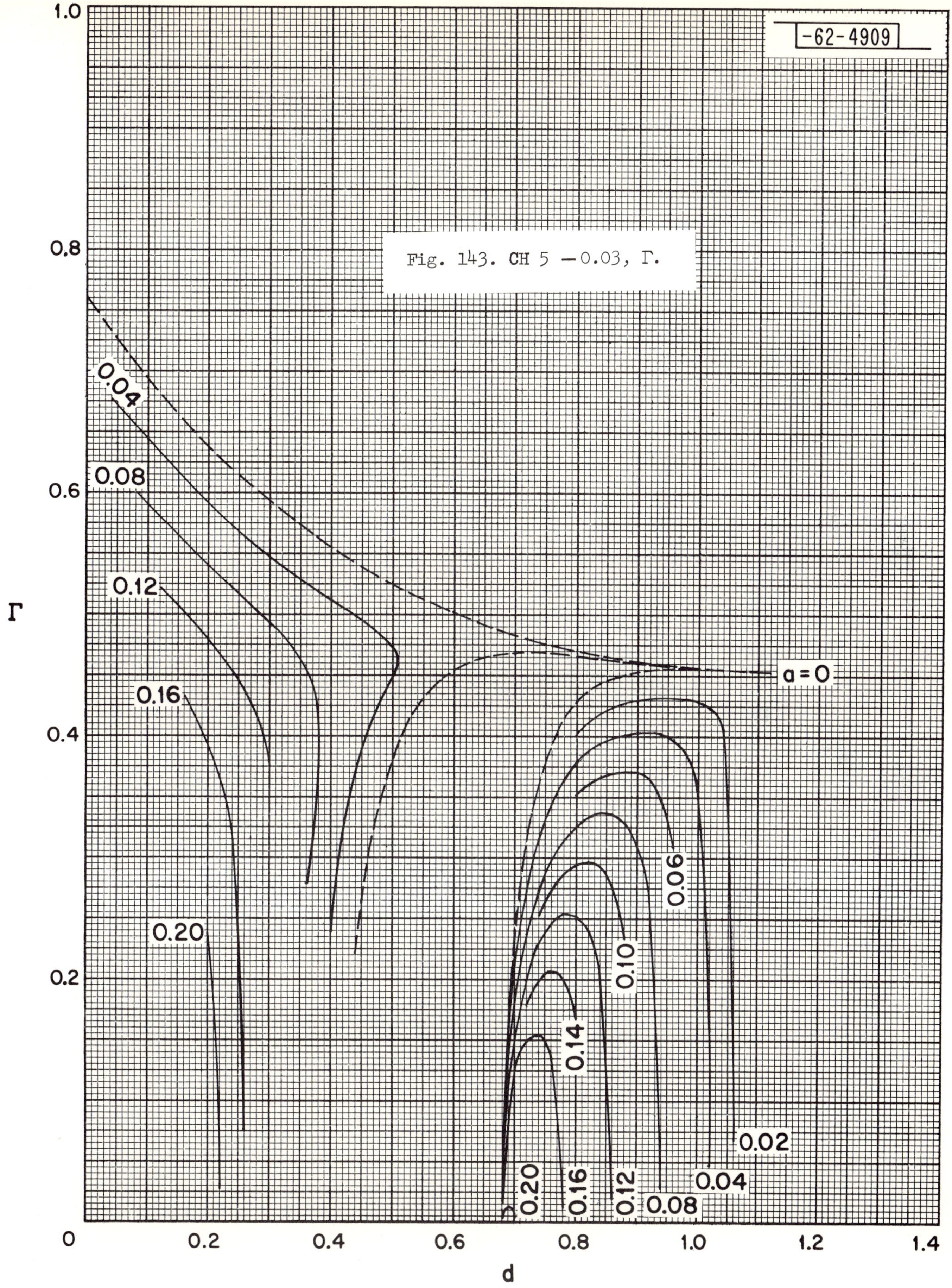

-62-4909
Fig. 143. CH 5 —0.03, Γ.
Γ
d
1.0
0.8
0.6
0.4
0.2
0
1.2
1.4
0.04
0.08
0.12
0.16
0.20
a = 0
0.02
0.04
0.06
0.08
0.10
0.12
0.14
0.16
0.20

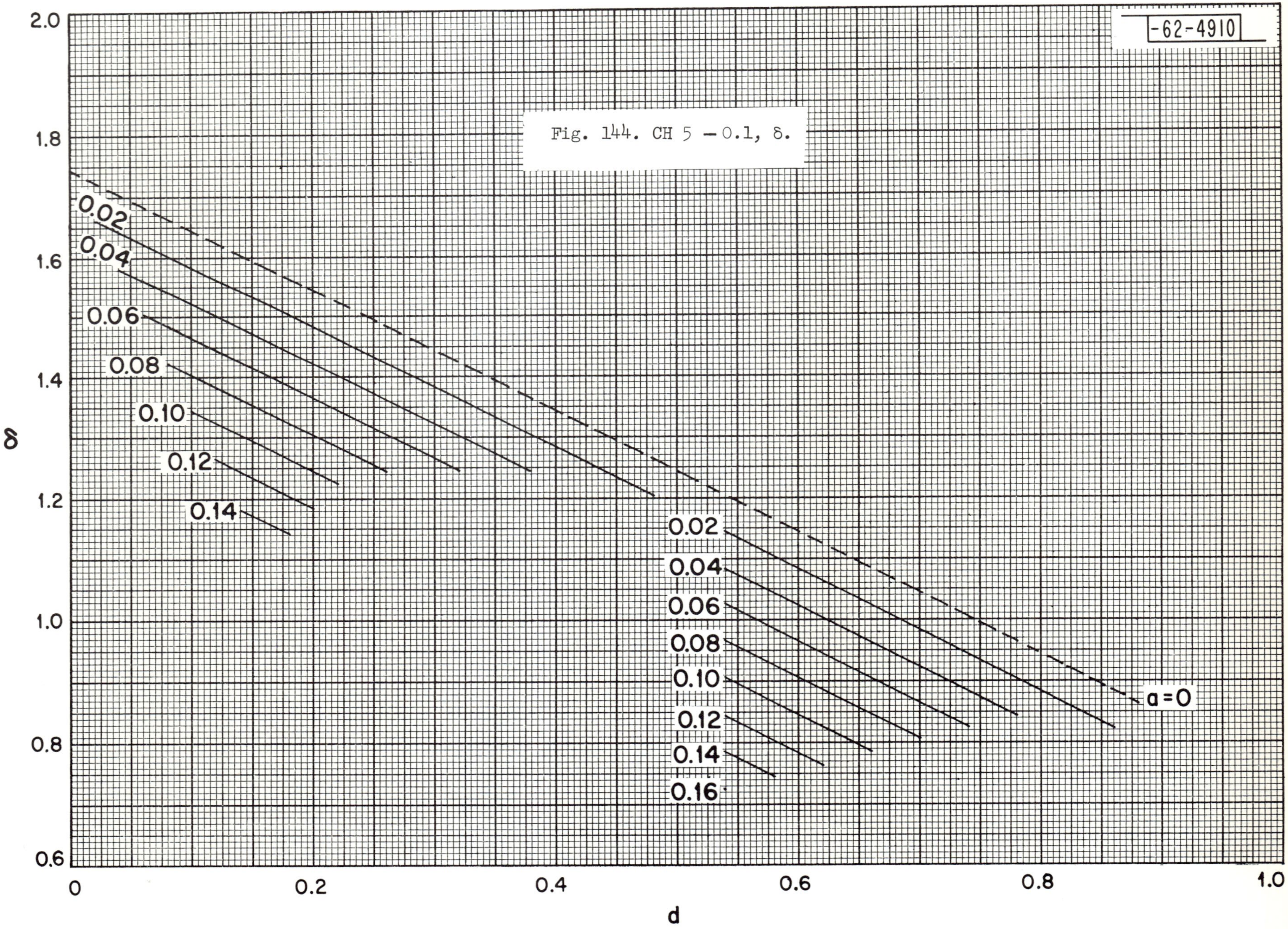


Fig. 144. CH 5 — 0.1, δ.

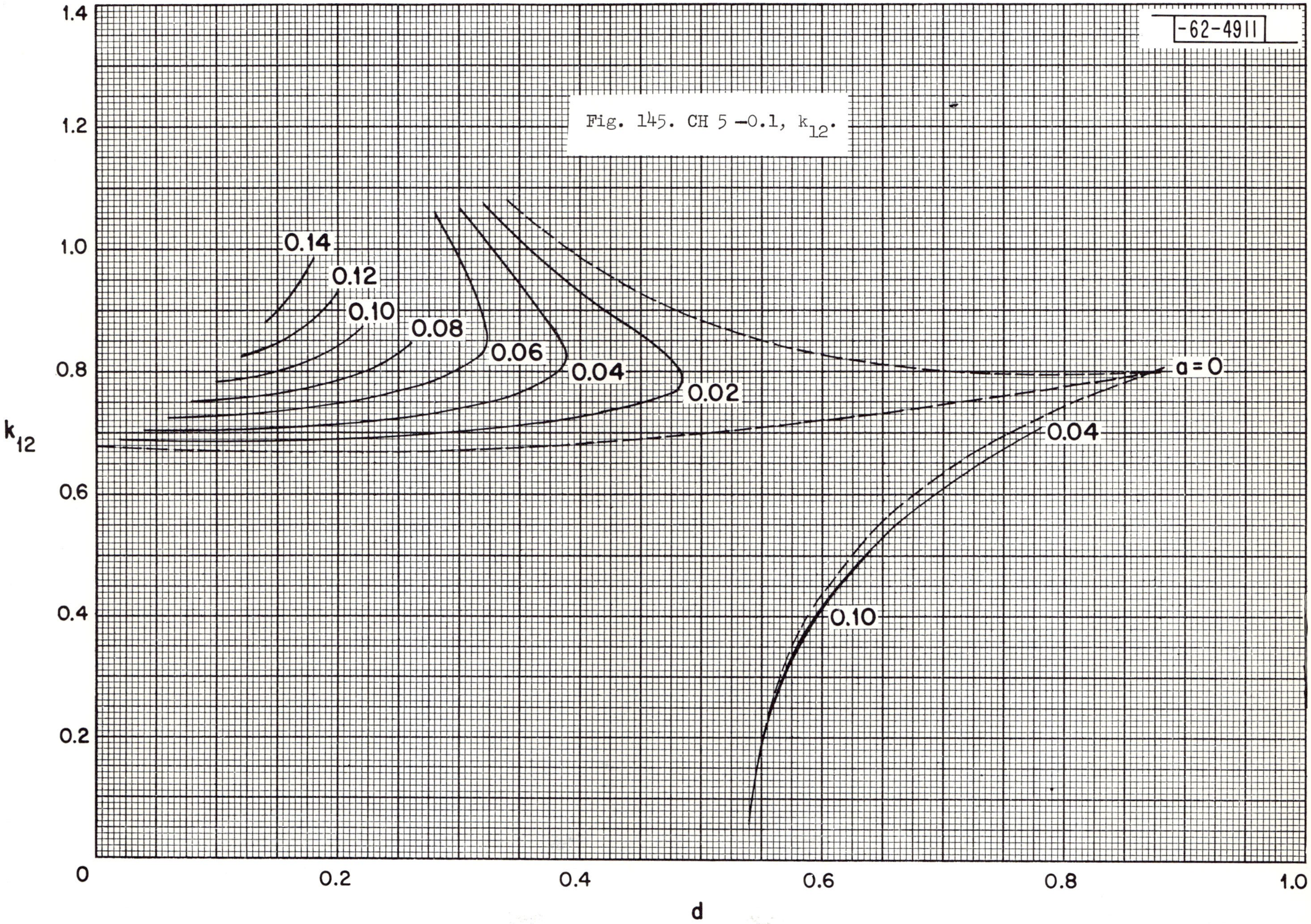


Fig. 145. CH 5 —0.1, $k_{12}$.

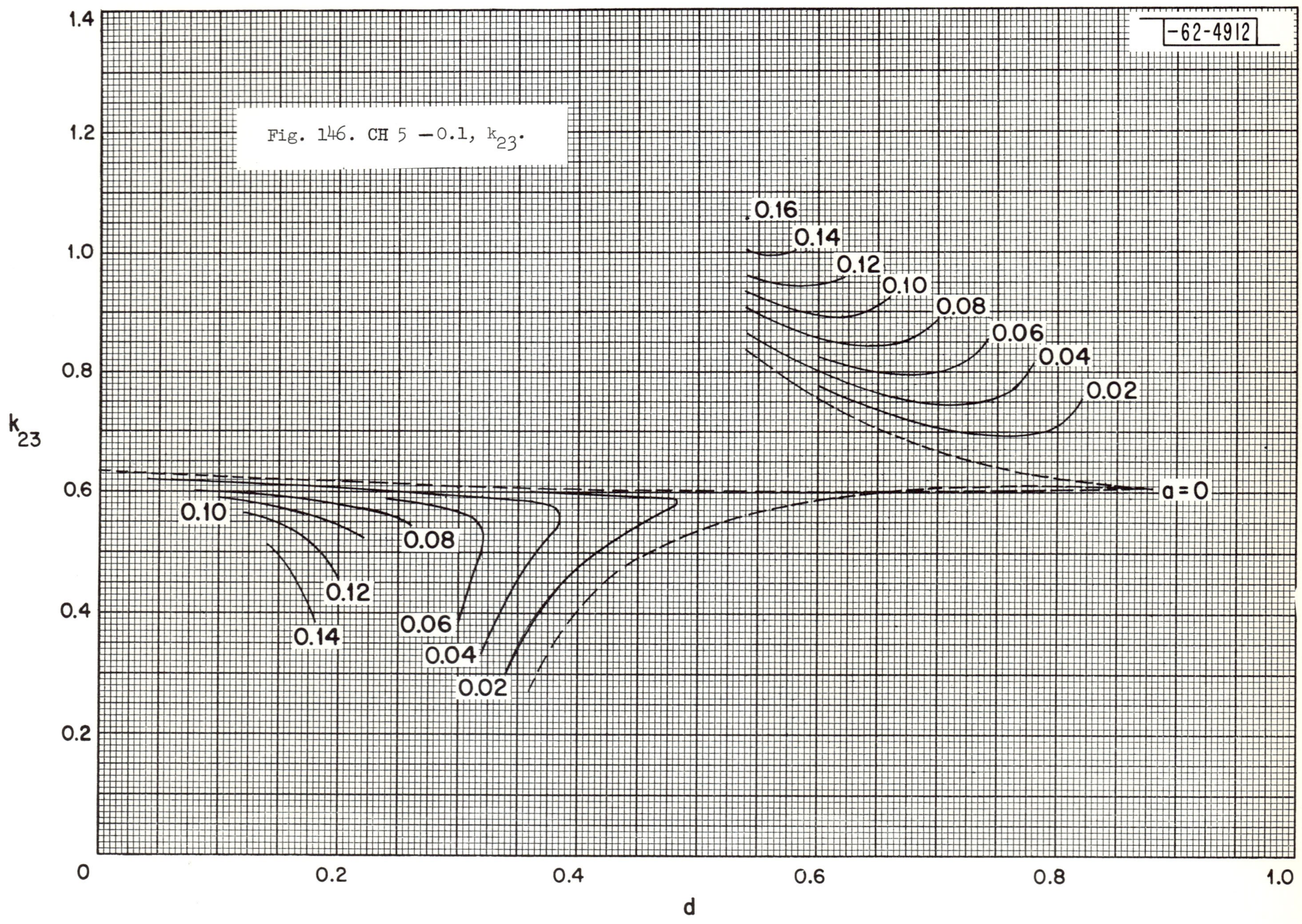


Fig. 146. CH 5 —0.1, $k_{23}$.

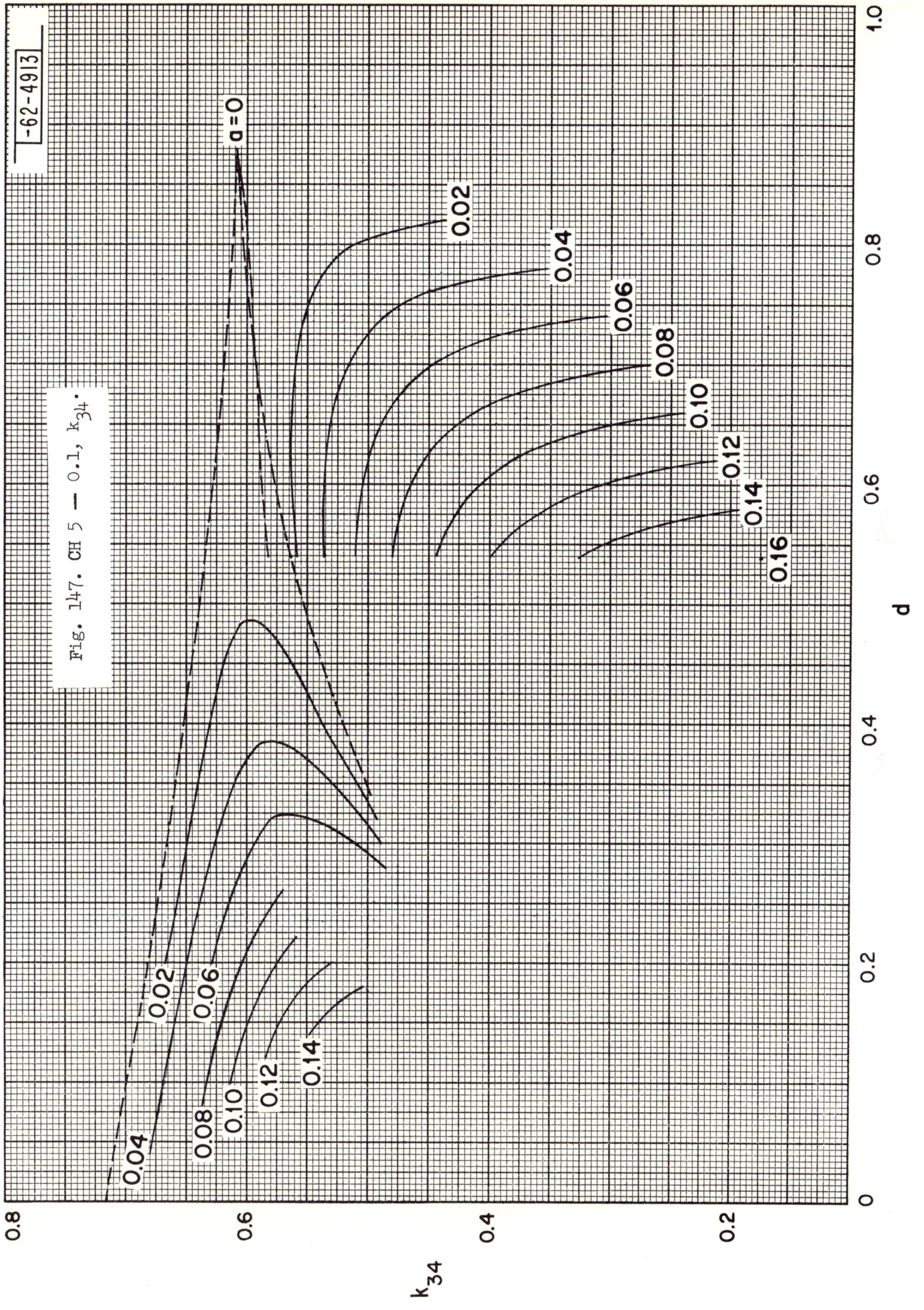


Fig. 147. CH 5 — 0.1, $k_{34}$.

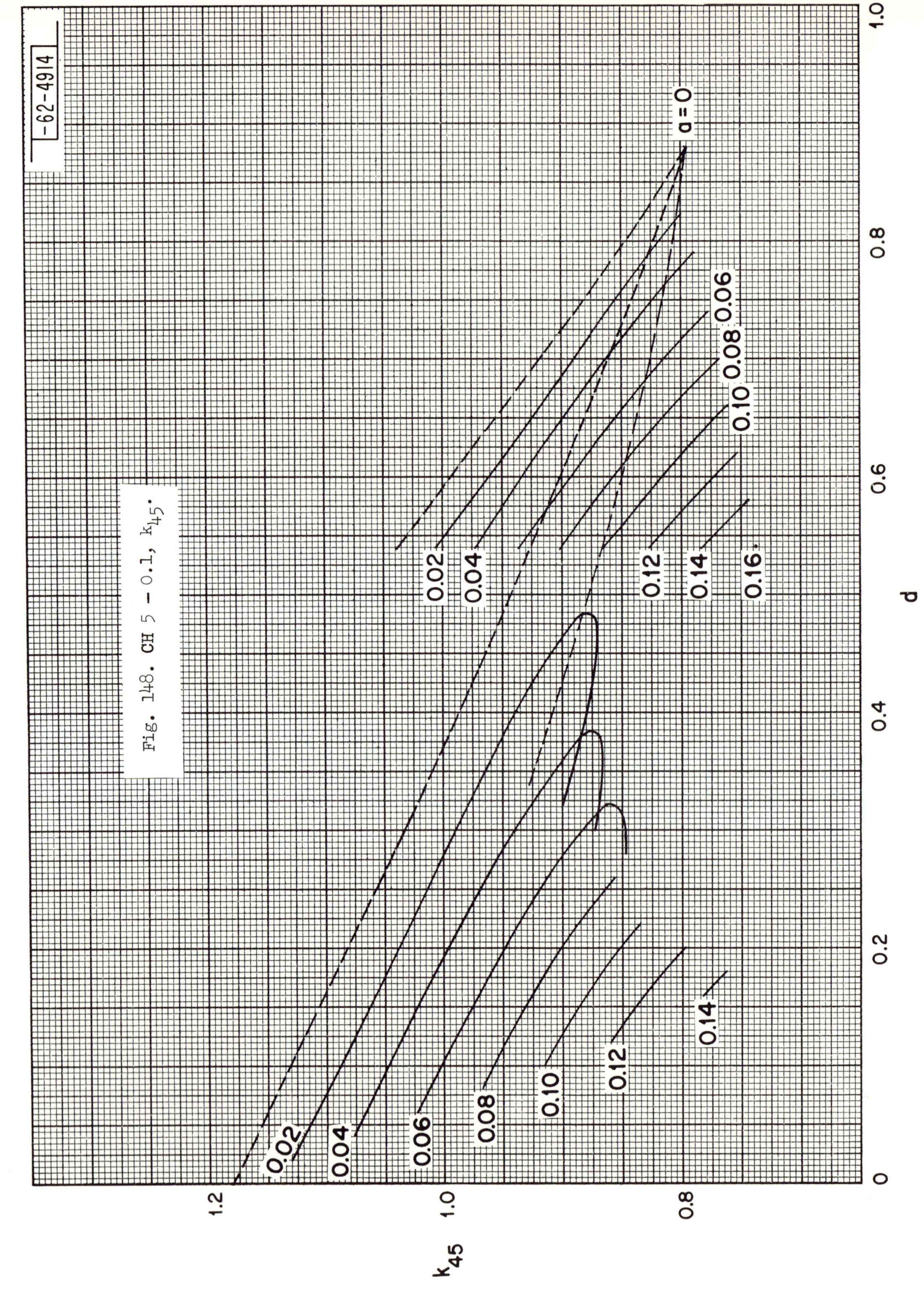


Fig. 148. CH 5 − 0.1, $k_{45}$.

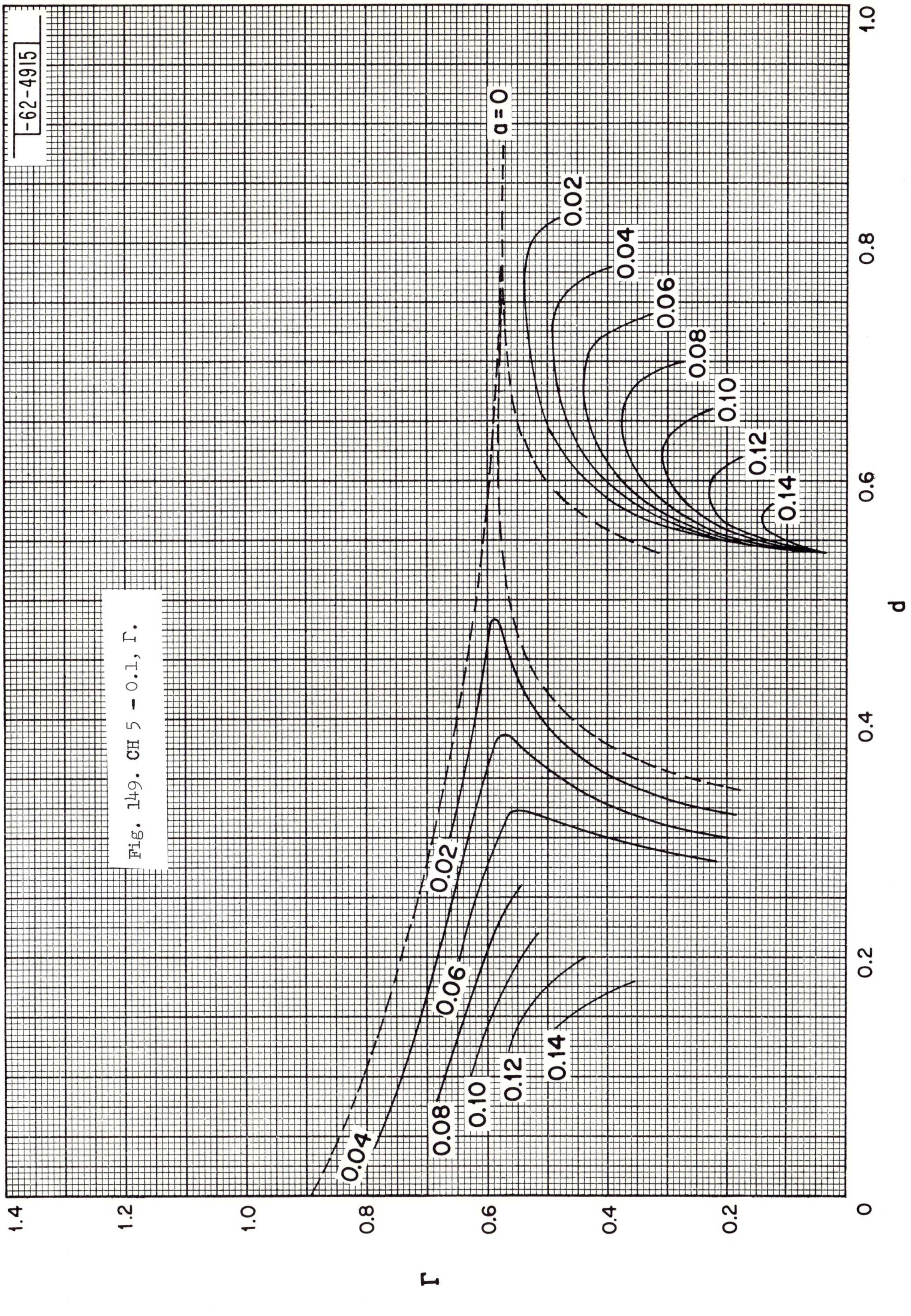


Fig. 149. CH 5 – 0.1, Γ.

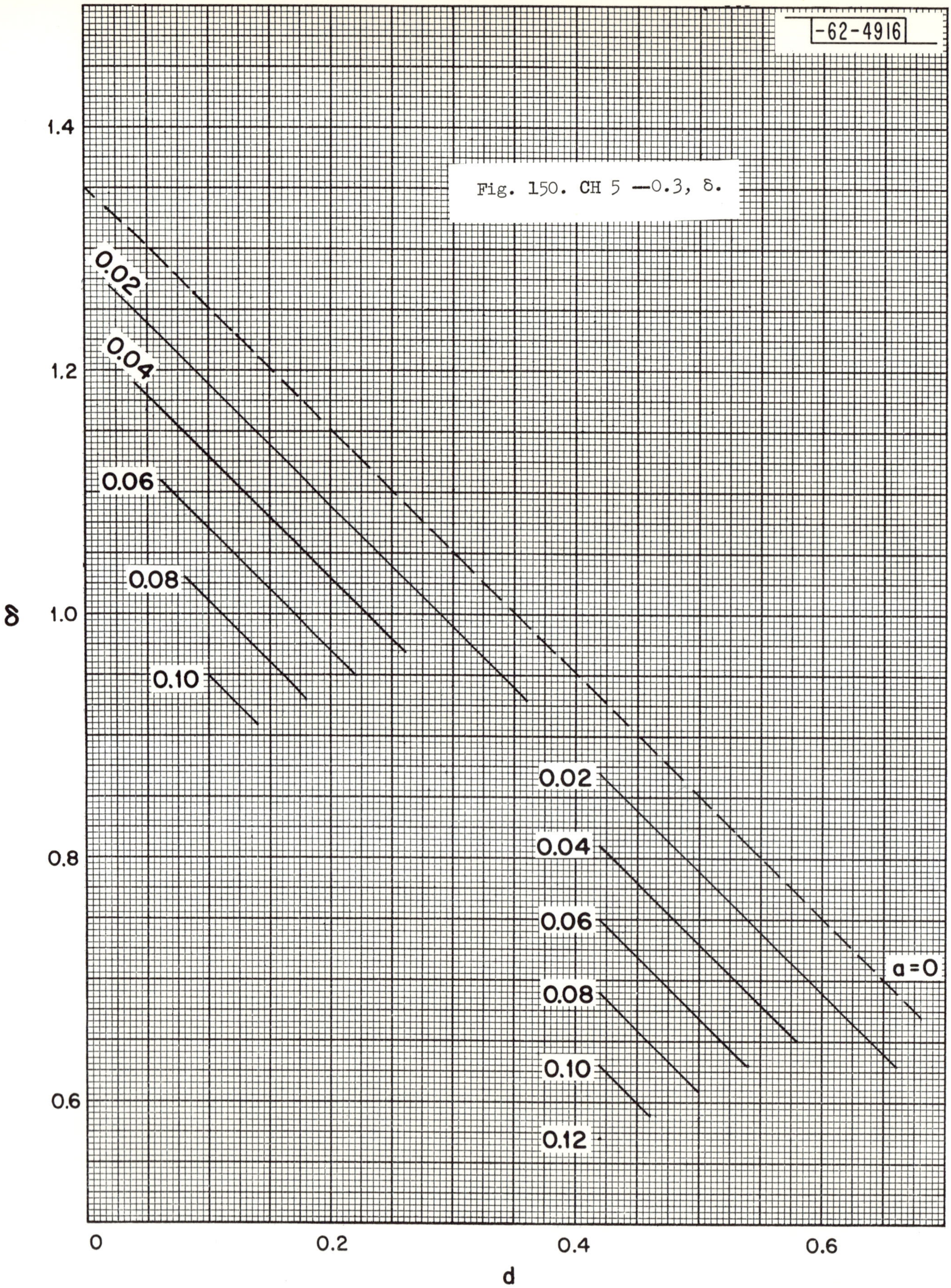

-62-4916
Fig. 150. CH 5 —0.3, δ.
1.4
1.2
1.0
0.8
0.6
δ
0.02
0.04
0.06
0.08
0.10
0.02
0.04
0.06
0.08
0.10
0.12
a = 0
0
0.2
0.4
0.6
d

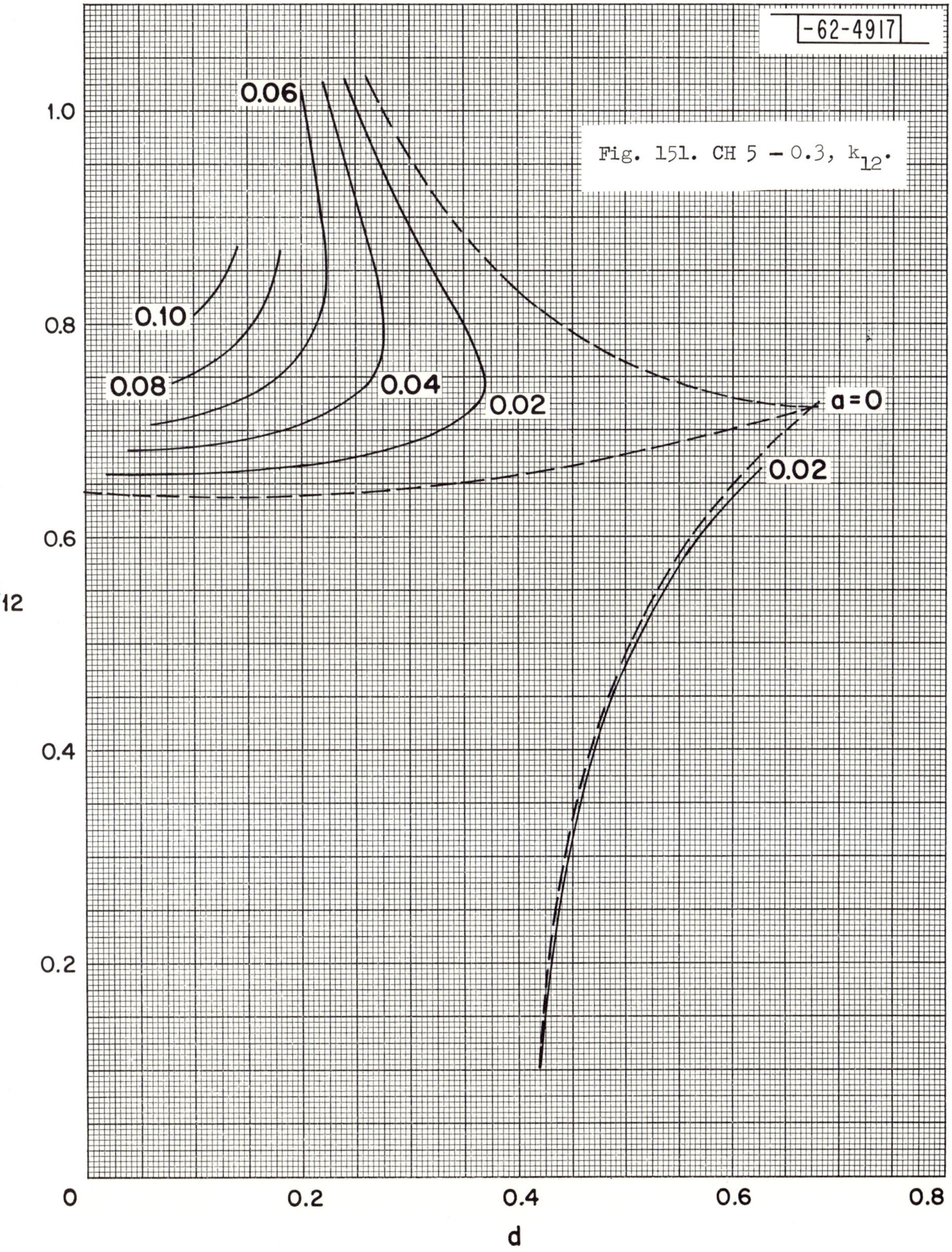


Fig. 151. CH 5 – 0.3, $k_{12}$.

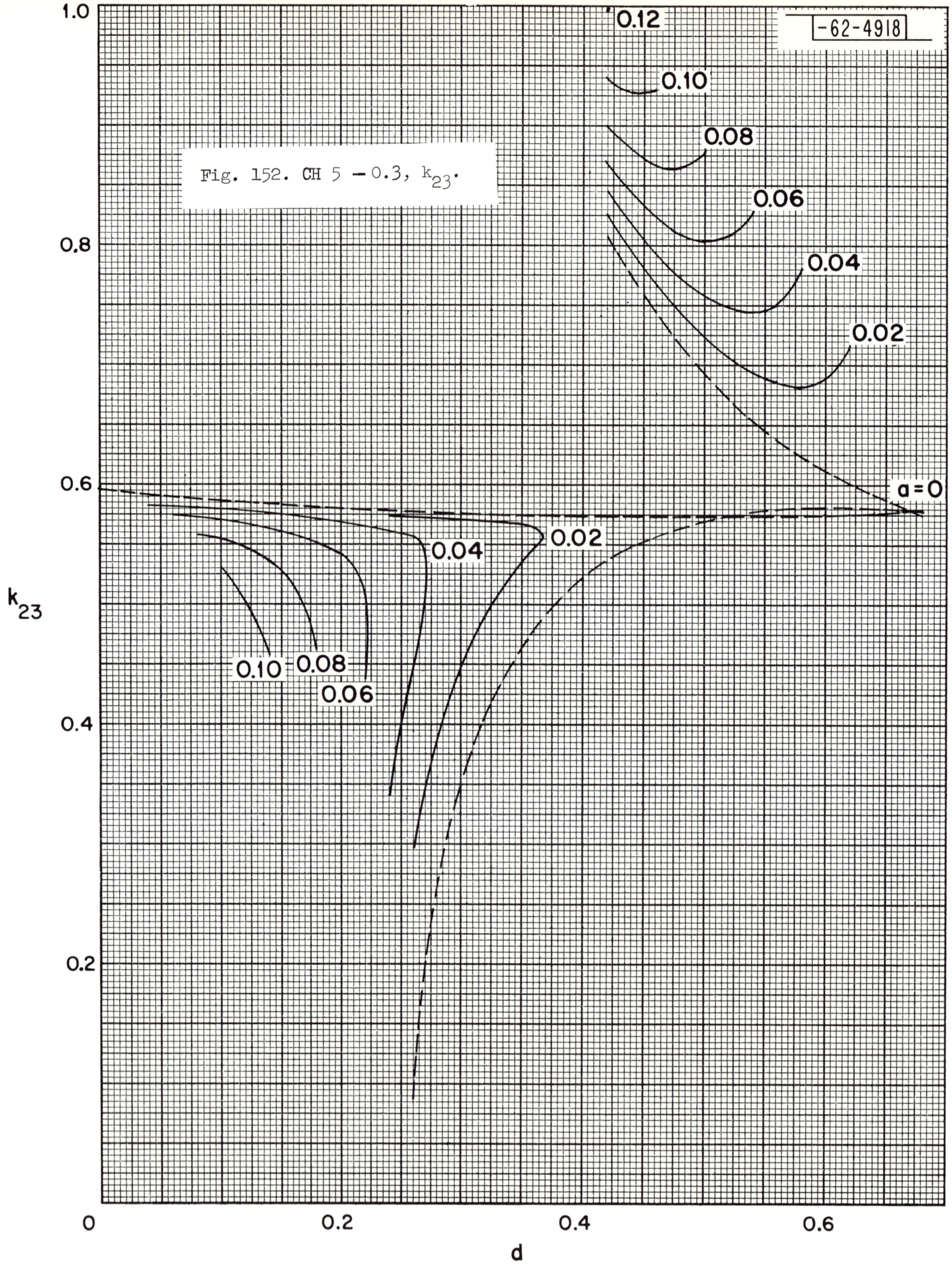
-62-4918
Fig. 152. CH 5 — 0.3, k23.
0.12
0.10
0.08
0.06
0.04
0.02
a=0
0.04
0.02
0.10
0.08
0.06
1.0
0.8
0.6
0.4
0.2
k23
0
0.2
0.4
0.6
d

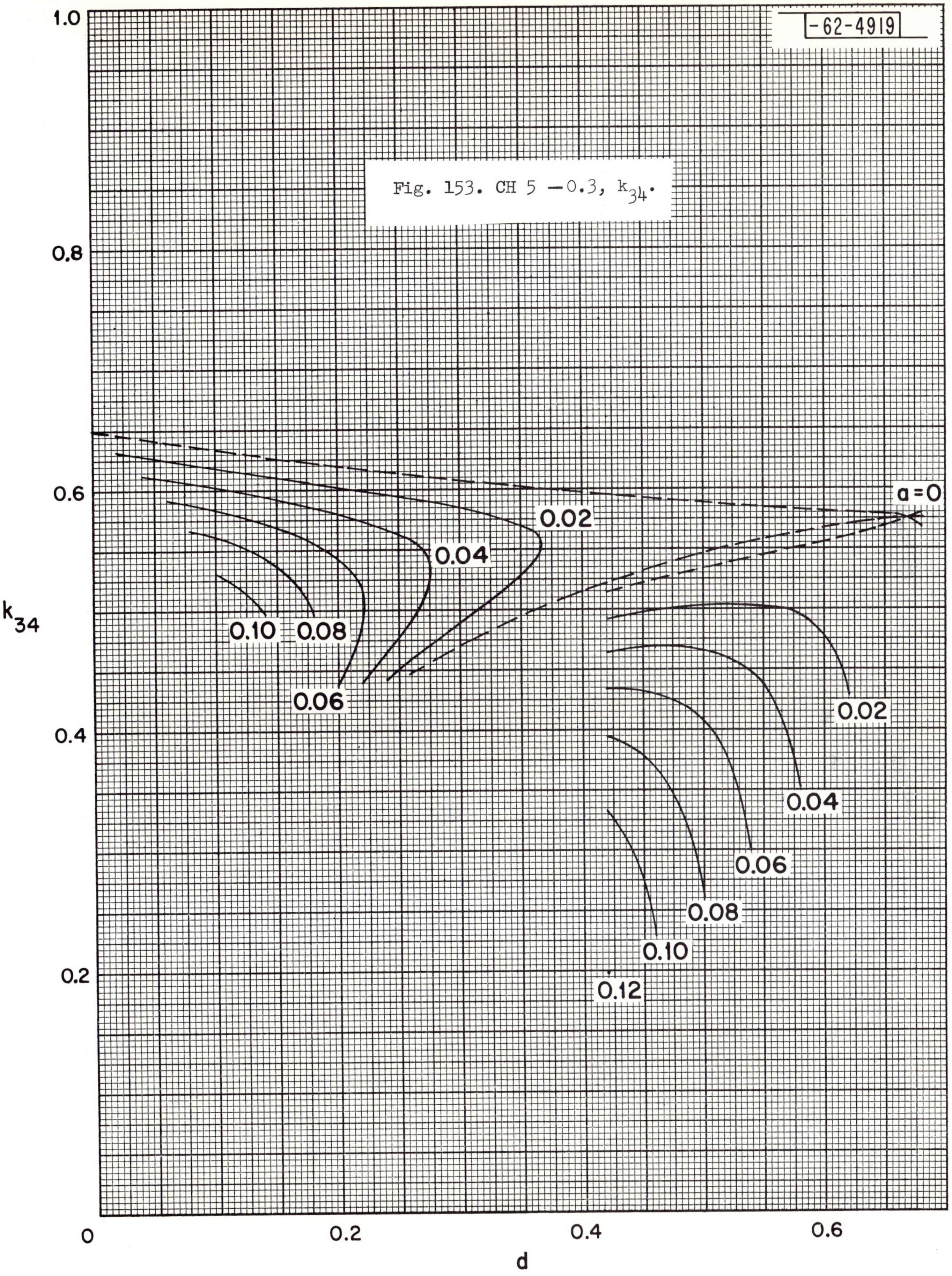


Fig. 153. CH 5 —0.3, $k_{34}$.

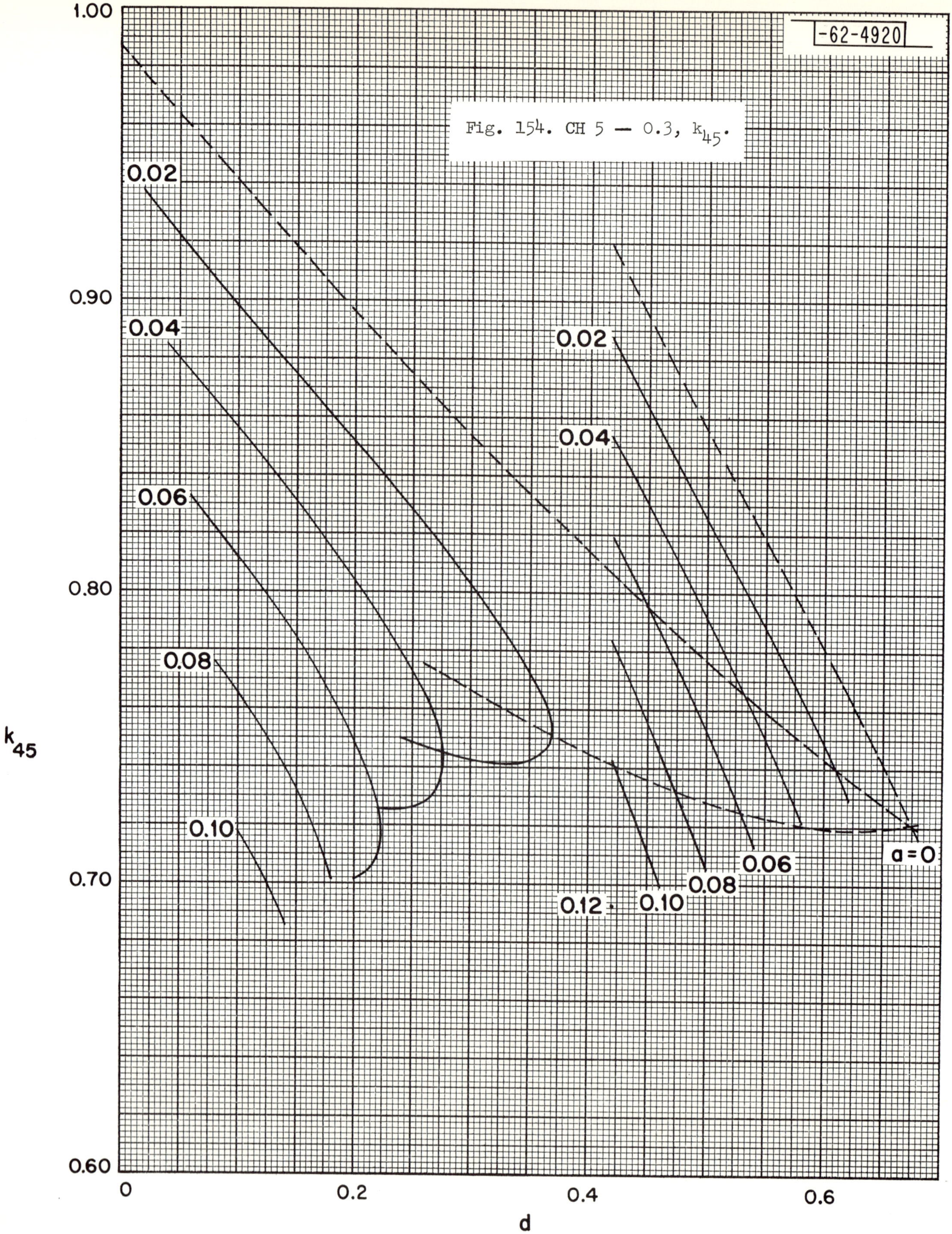

Fig. 154. CH 5 — 0.3, $k_{45}$.

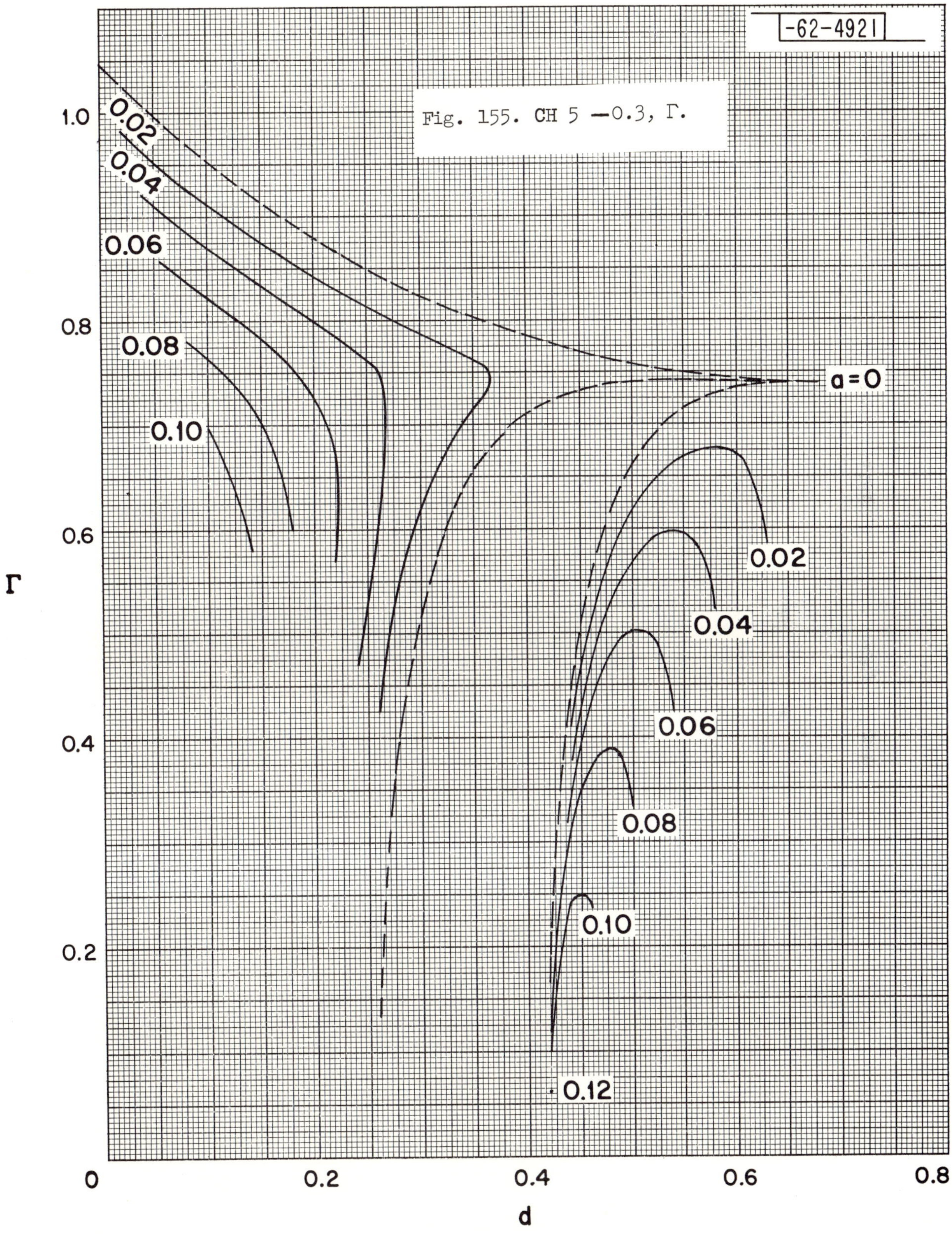


Fig. 155. CH 5 —0.3, Γ.

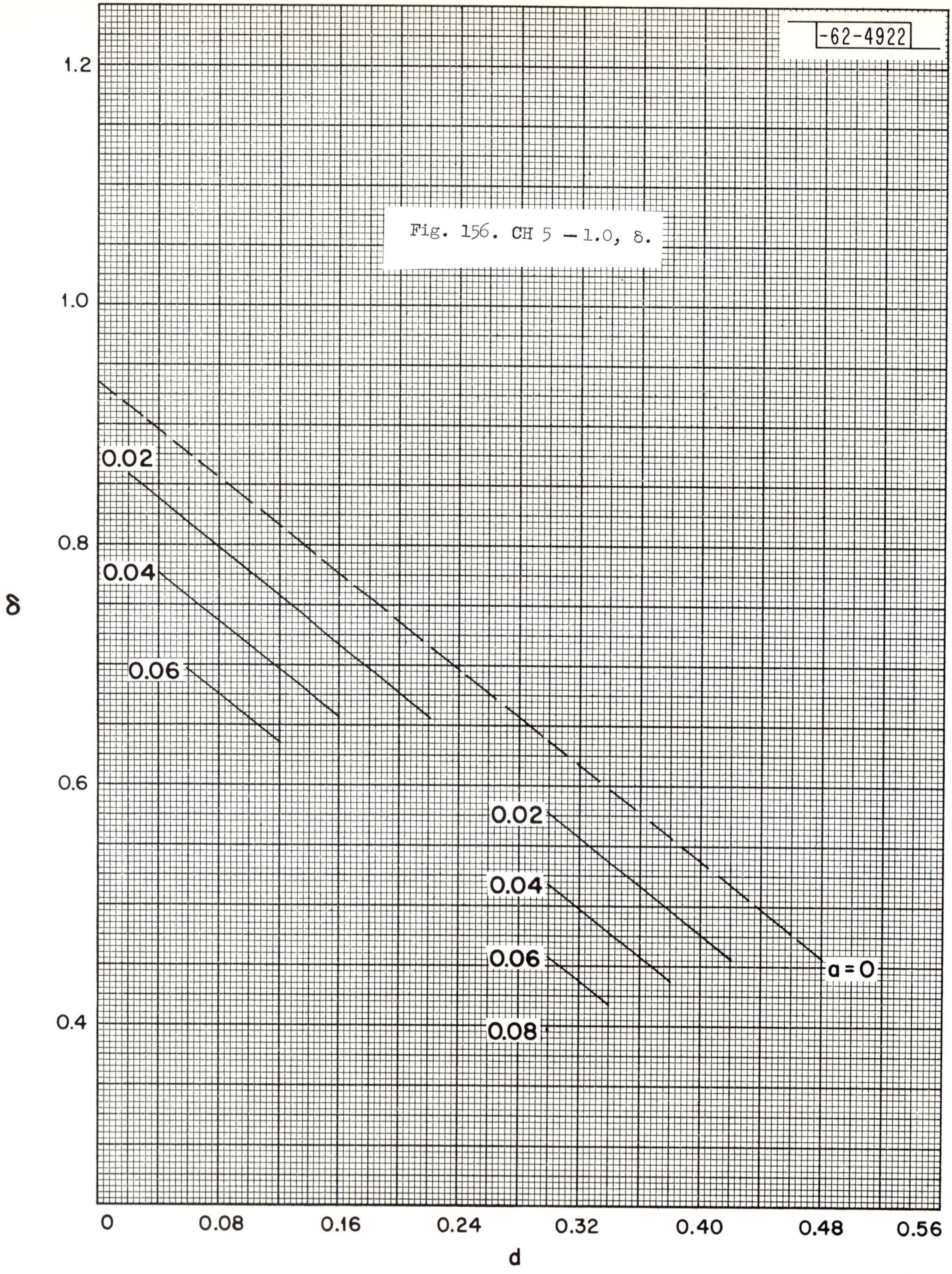
-62-4922
Fig. 156. CH 5 — 1.0, δ.
1.2
1.0
0.8
0.6
0.4
δ
0.02
0.04
0.06
0.02
0.04
0.06
0.08
a = 0
0
0.08
0.16
0.24
0.32
0.40
0.48
0.56
d

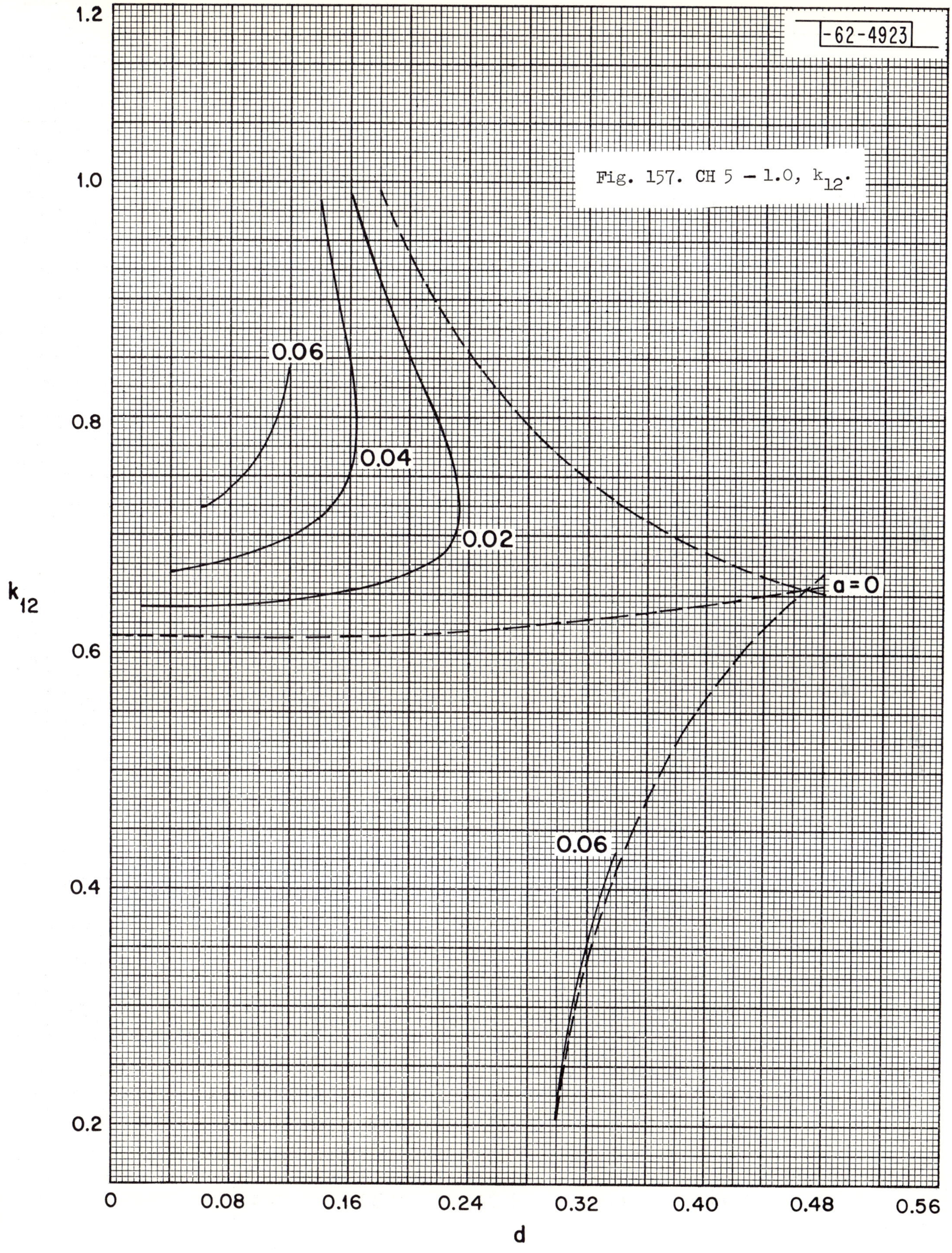


Fig. 157. CH 5 — 1.0, $k_{12}$.

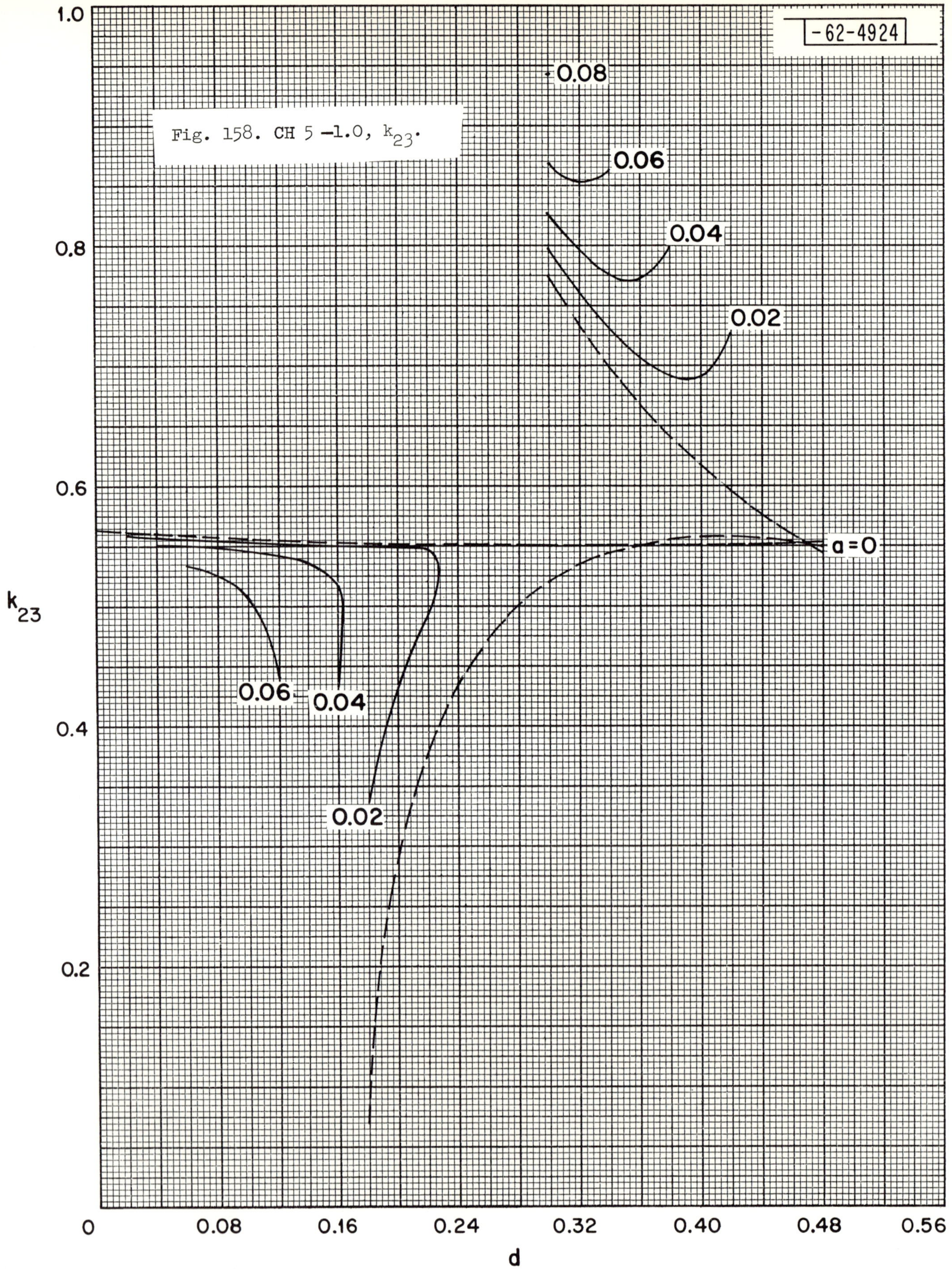

-62-4924
Fig. 158. CH 5 -1.0, $k_{23}$.
0.08
0.06
0.04
0.02
a=0
0.06
0.04
0.02
$k_{23}$
d
1.0
0.8
0.6
0.4
0.2
0
0.08
0.16
0.24
0.32
0.40
0.48
0.56

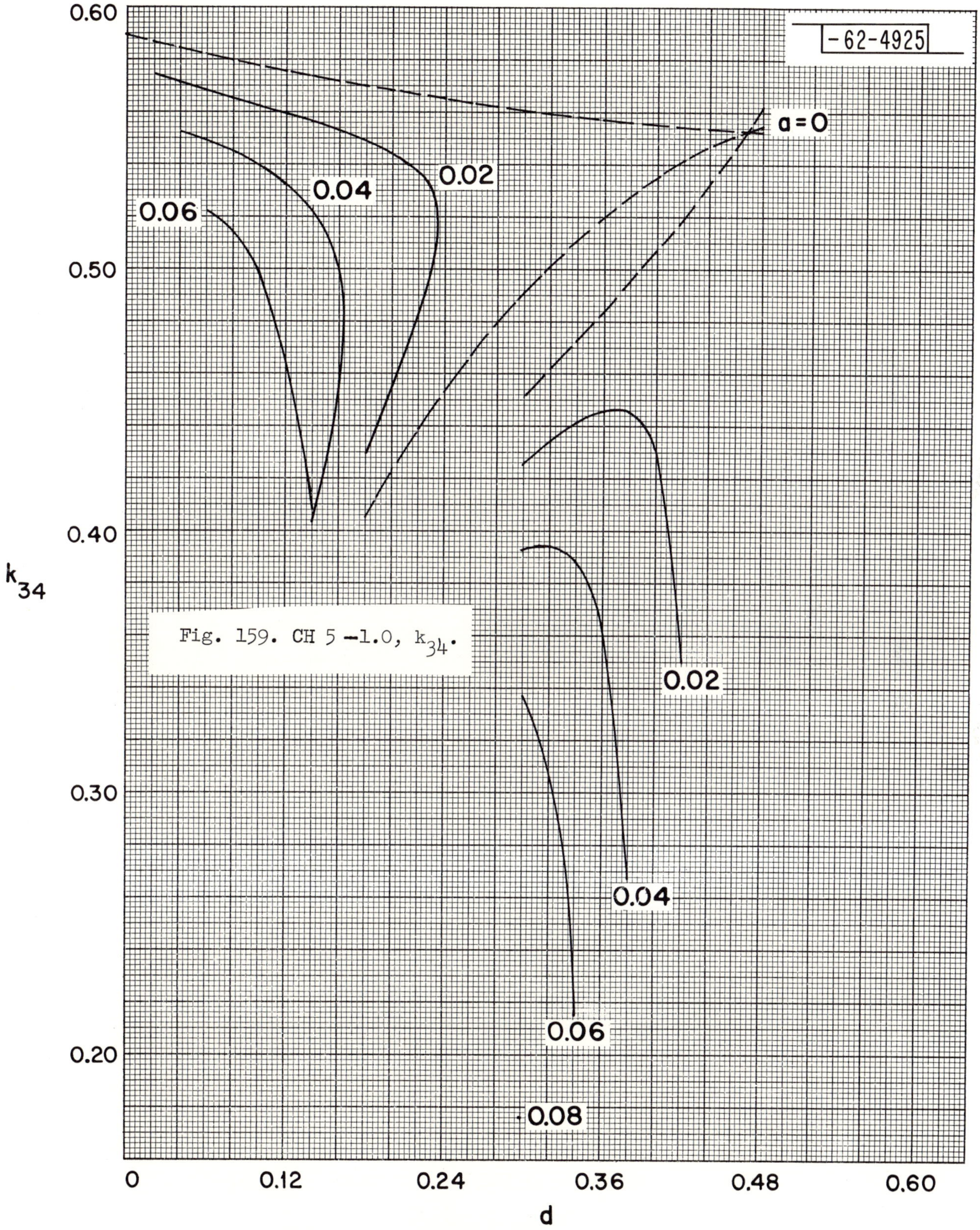


Fig. 159. CH 5 —1.0, $k_{34}$.

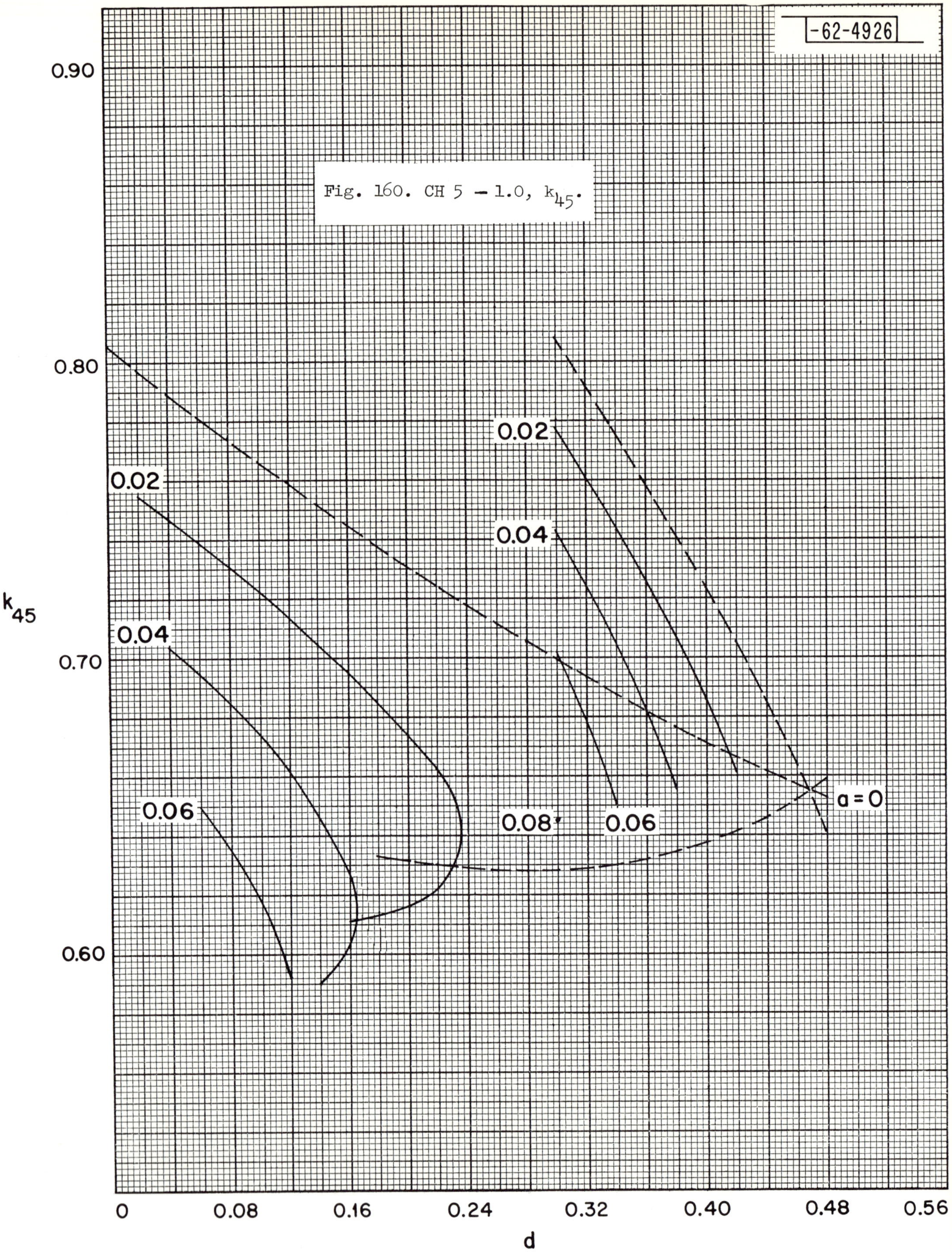


Fig. 160. CH 5 — 1.0, $k_{45}$.

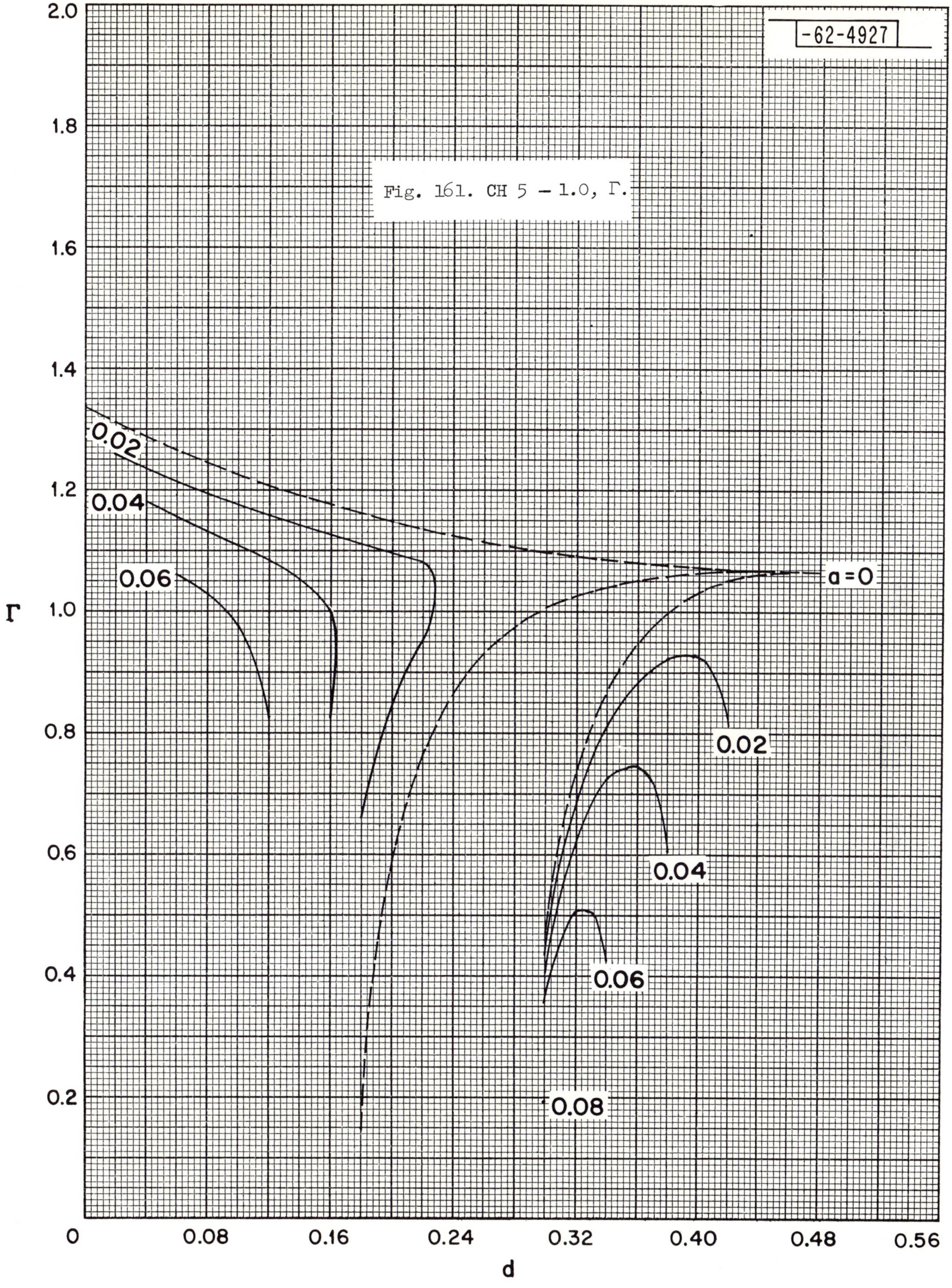

-62-4927
Fig. 161. CH 5 – 1.0, Γ.
2.0
1.8
1.6
1.4
1.2
1.0
0.8
0.6
0.4
0.2
Γ
0.02
0.04
0.06
a=0
0.02
0.04
0.06
0.08
0
0.08
0.16
0.24
0.32
0.40
0.48
0.56
d

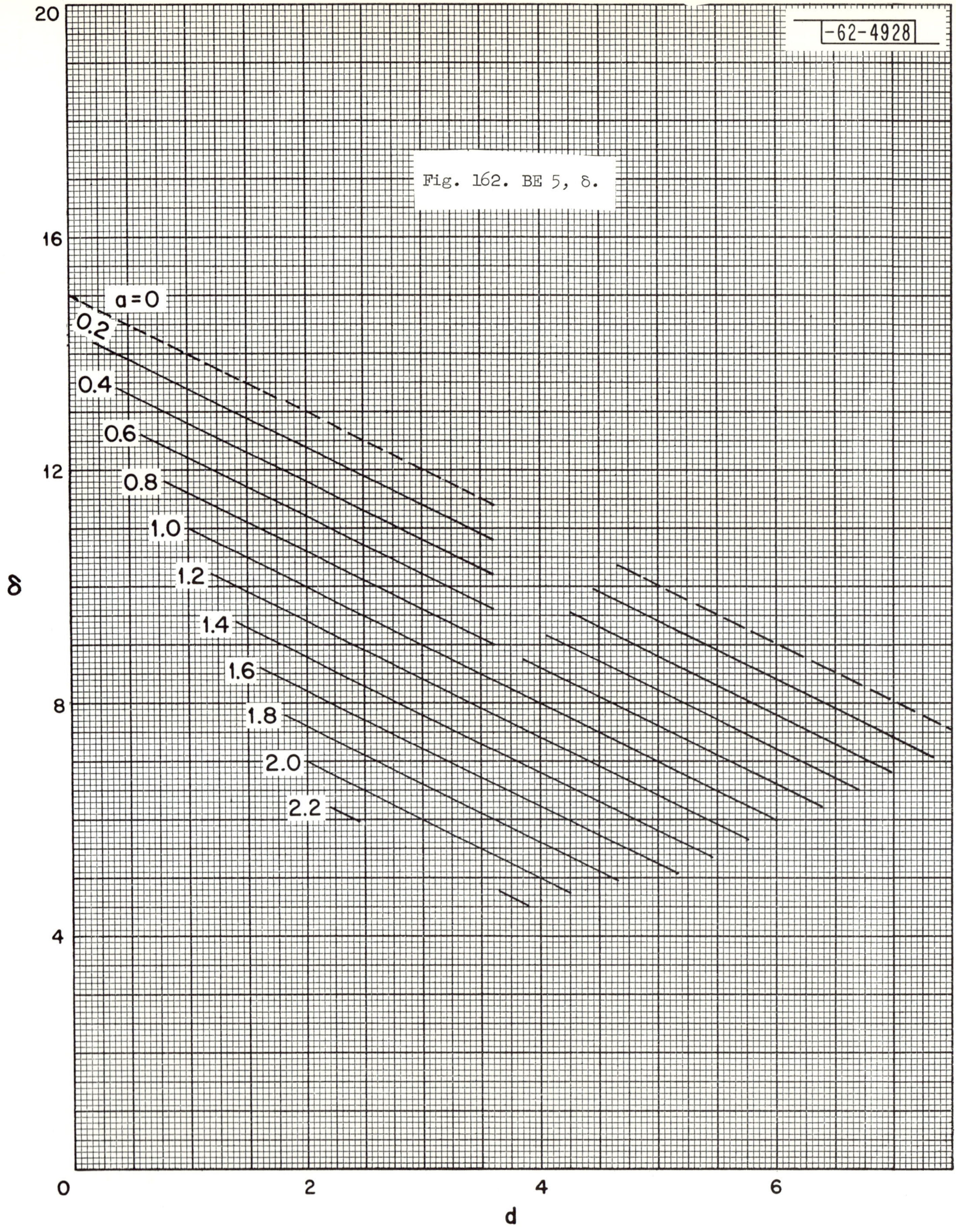
-62-4928
Fig. 162. BE 5, δ.
a=0
0.2
0.4
0.6
0.8
1.0
1.2
1.4
1.6
1.8
2.0
2.2
δ
d
20
16
12
8
4
0
2
4
6

Fig. 163. BE 5, $k_{12}$.

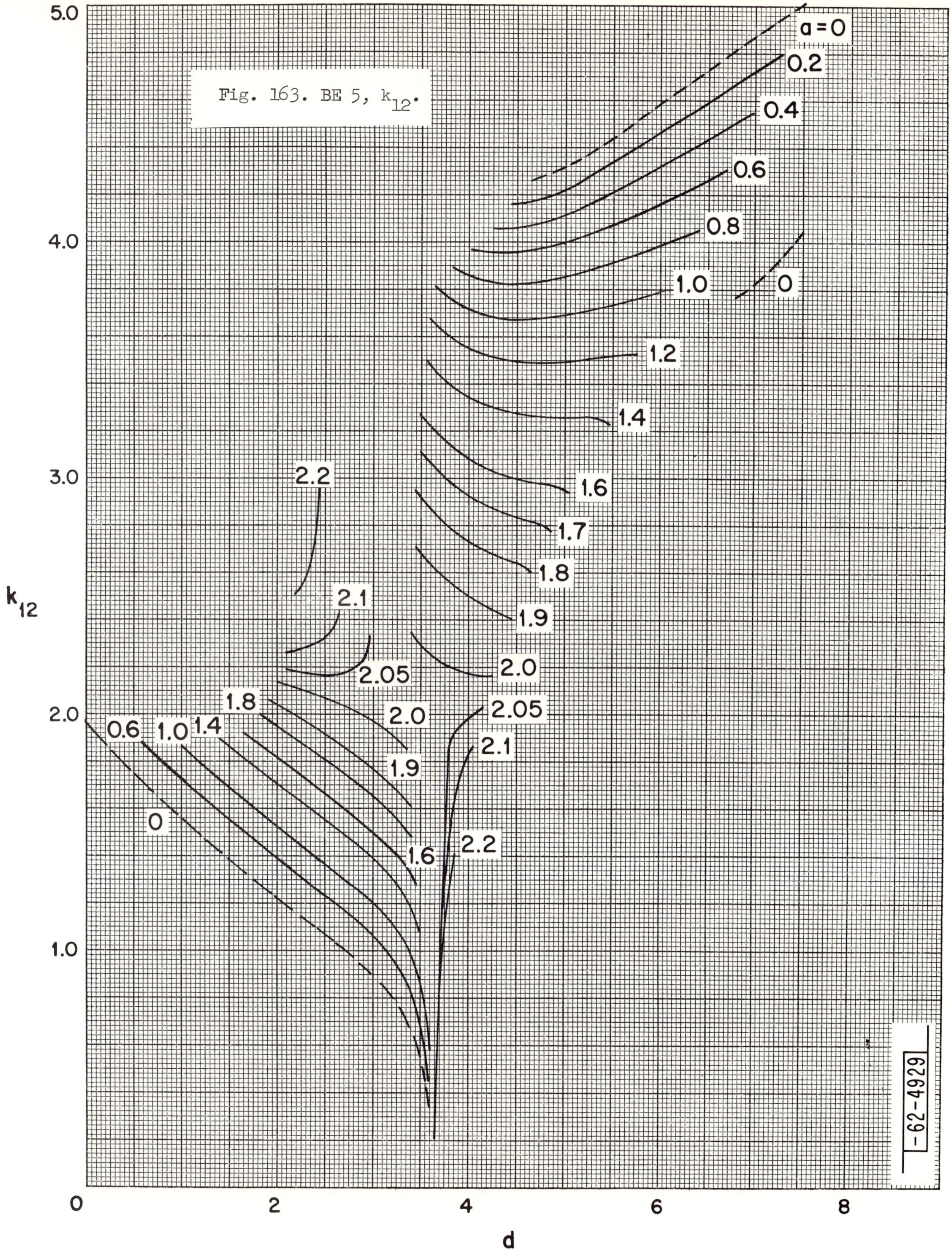

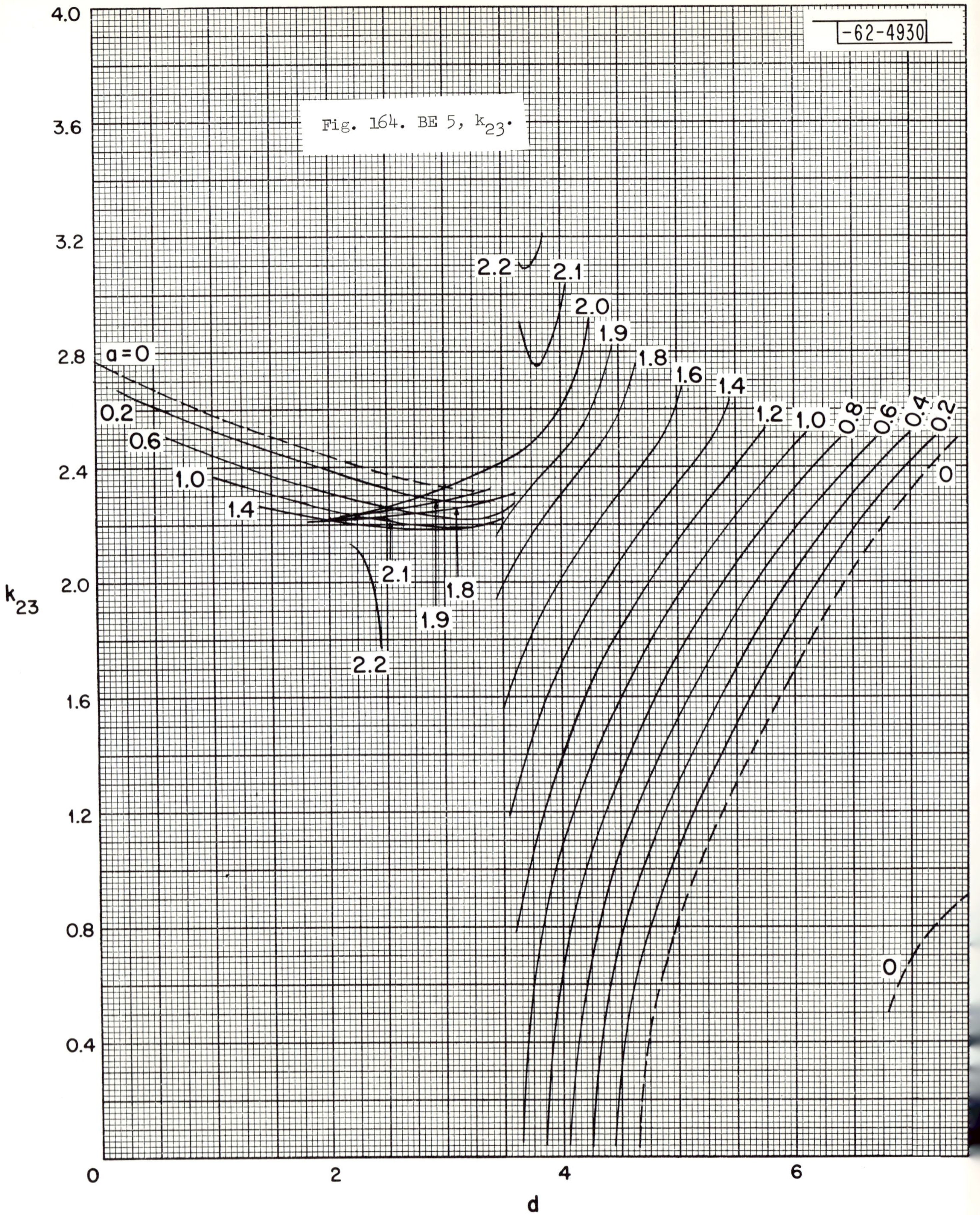

Fig. 164. BE 5, $k_{23}$.

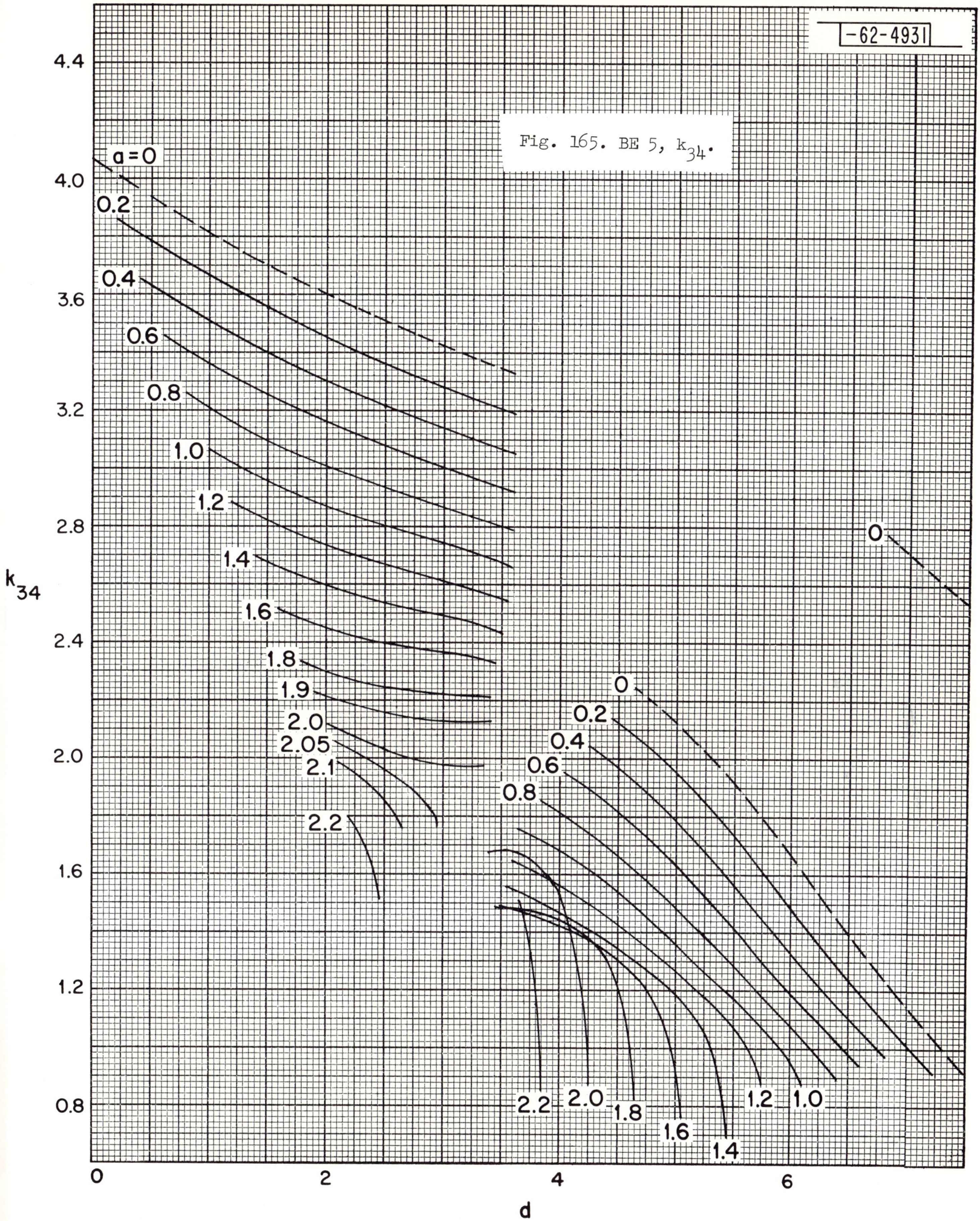

-62-4931
Fig. 165. BE 5, $k_{34}$.
a=0
0.2
0.4
0.6
0.8
1.0
1.2
1.4
1.6
1.8
1.9
2.0
2.05
2.1
2.2
0
0.2
0.4
0.6
0.8
2.2
2.0
1.8
1.6
1.4
1.2
1.0
$k_{34}$
4.4
4.0
3.6
3.2
2.8
2.4
2.0
1.6
1.2
0.8
0
2
4
6
d

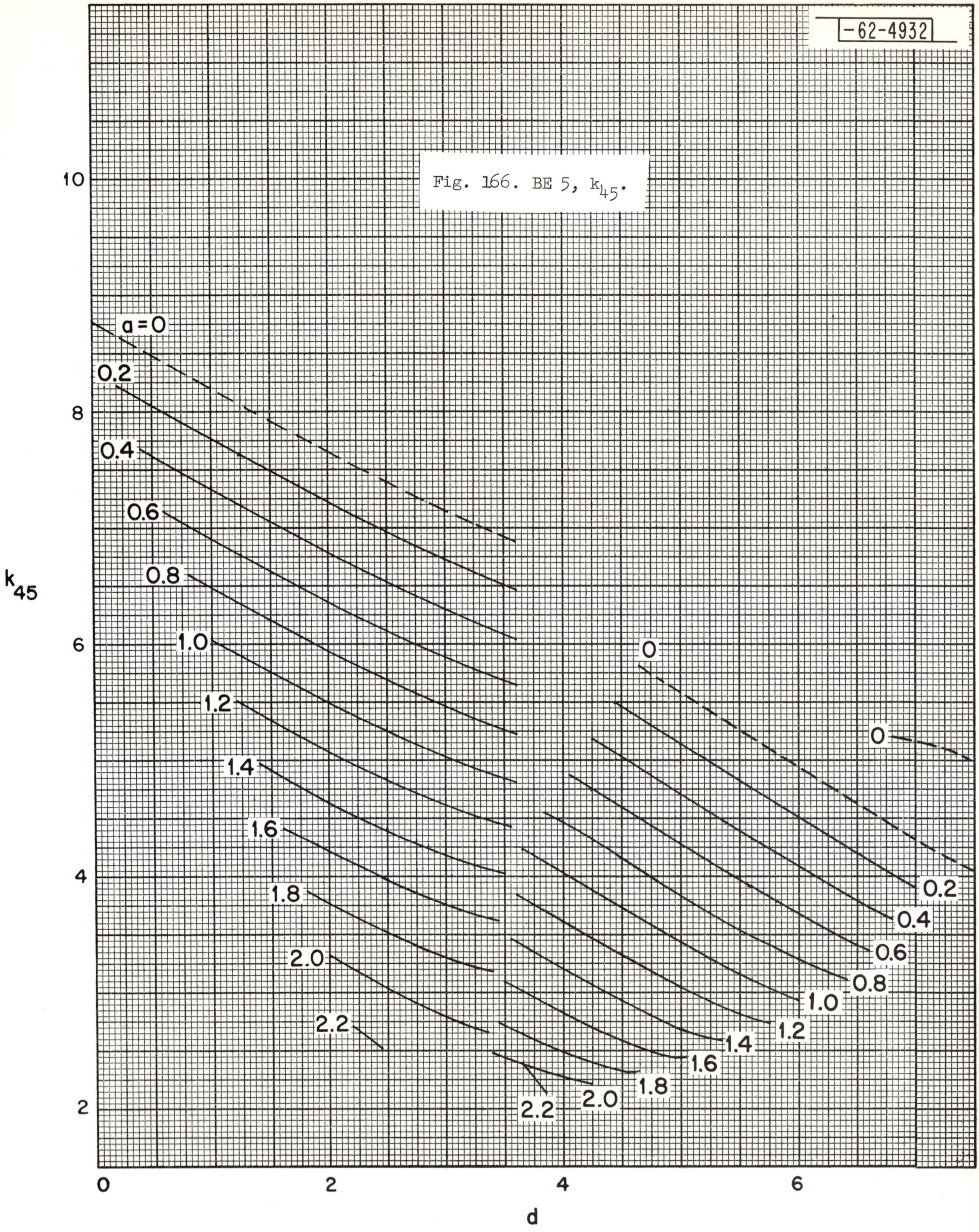


Fig. 166. BE 5, $k_{45}$.

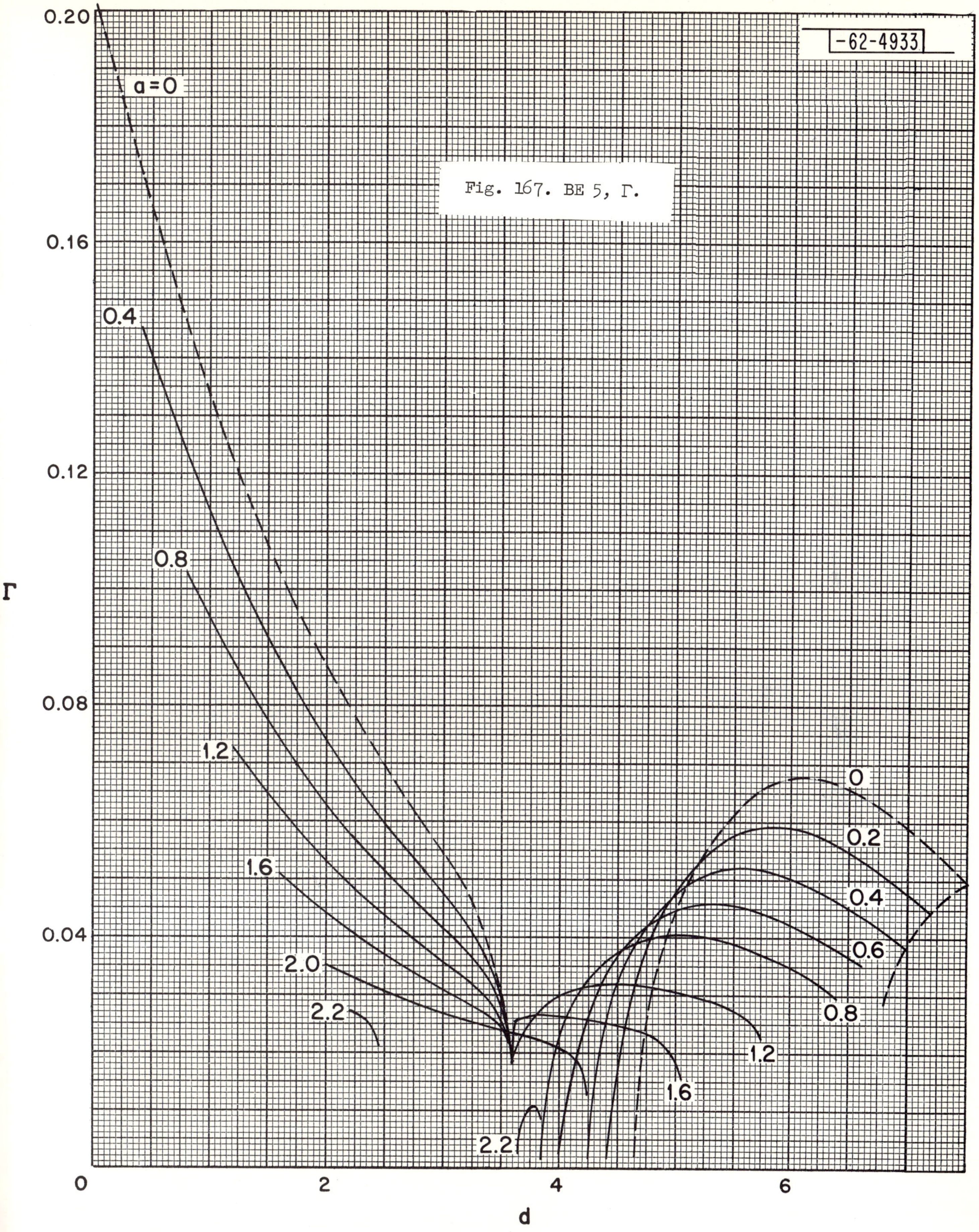
-62-4933
Fig. 167. BE 5, Γ.
a=0
0.4
0.8
1.2
1.6
2.0
2.2
0
0.2
0.4
0.6
0.8
1.2
1.6
2.2
0.20
0.16
0.12
0.08
0.04
0
2
4
6
Γ
d

## APPENDIX A
## ALIGNMENT PROCEDURES

The alignment of narrow bandpass filters is fully covered by Dishal,[20] so only a summary will be given here. The object of the alignment is to tune the self admittance (impedance) of each node (loop) to the specified center frequency. The coupling elements should be considered as fixed, so one of the other two elements at each node (or in each loop) must be variable and the other fixed.

A convenient alignment procedure, requiring no internal measurements on the filter, is as follows (the coupled node circuit of Fig. A-1(a) with variable capacitors is used as a specific illustration). Apply a sine wave of frequency $f_0$ from a high impedance source to port A, put a high impedance and low capacity voltmeter between node 1 and ground, and short circuit node 2 to ground. The impedances of the source and voltmeter should be high enough not to broaden appreciably the resonance of $L_1$ and $C_1$; adjustment of the resonant frequency should be to the order of 1/2 percent of the filter bandwidth. Adjust $C_1$ so that the voltage $E_1$ is maximum. The self admittance of node 1 is now tuned properly. Next remove the short circuit from node 2 and short circuit node 3 to ground. Adjust $C_2$ so that $E_1$ is a minimum; now the self admittance of node 2 is properly tuned. Remove the short circuit from node 3 and short circuit node 4 to ground. Adjust $C_3$ so that $E_1$ is a maximum. Continue in this way until the end of the network is reached, and when adjusting $C_n$ be sure that $R_\ell$ is disconnected. After $C_n$ is adjusted, the network is aligned except for the effect of capacitance of the voltmeter across $C_1$. After the voltmeter is removed, the final adjustment of $C_1$ can be made by exciting the properly terminated network from a sweep generator and adjusting $C_1$ for best response. Usually the response of the filter is now satisfactory, but sometimes it is necessary to make further adjustments of the other C's while observing the response with the sweep generator.

Sometimes in the alignment the adjustment of the C's near the output cannot be made accurately enough by working from the input; it is then advantageous to work from both ends toward the center.

Alignment of the other narrow bandpass circuits is done in an analogous manner. For the coupled loop circuit of Fig. A-1(b), apply a sine wave of frequency $f_0$ from a low impedance source to the input port and put a low impedance current probe in link 1. Open link 2 and adjust one of the reactances of link 1 for maximum current in link 1. Next close link 2, open link 3 and adjust a reactance in link 2 for minimum current in link 1. Continue in this manner, alternately adjusting for maximum and minimum current, until all loops are tuned. Compensation for the inductance of the current probe and any other further trimming that is necessary can be done as before using a sweep generator.

The alignment of the mixed circuit in Fig. A-1(c) should require no explanation.

The general idea in the alignment of any of the narrow bandpass filters is to obtain coupling reactances that are within 3 percent of the calculated value and the remaining elements that are within 5 percent of the calculated value. Then tune each node or loop so that its tuning error is less than 1/2 percent of the bandwidth of the filter.

The problem of aligning the lowpass filters of Fig. A-2 is trivial. Components whose value is in error by less than 5 percent of the computed value will satisfy all but the most fussy. If one wants to be fussy a 3 percent tolerance is sufficient.

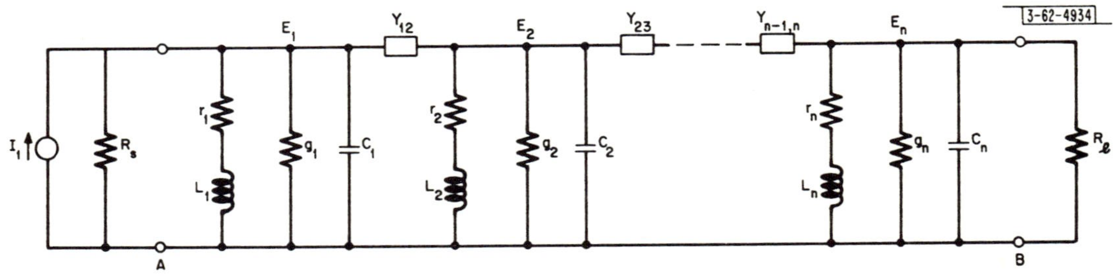


(a) Reactance-coupled parallel tuned circuits. With ports A and B open, self-admittance of each node is resonant at $\omega_0 = 2\pi f_0$.

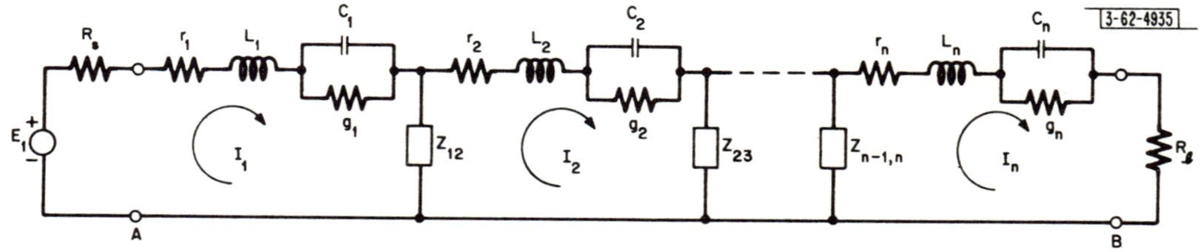


(b) Reactance-coupled series tuned circuits. With ports A and B short circuited, self-impedance of each loop is resonant at $\omega_0 = 2\pi f_0$.

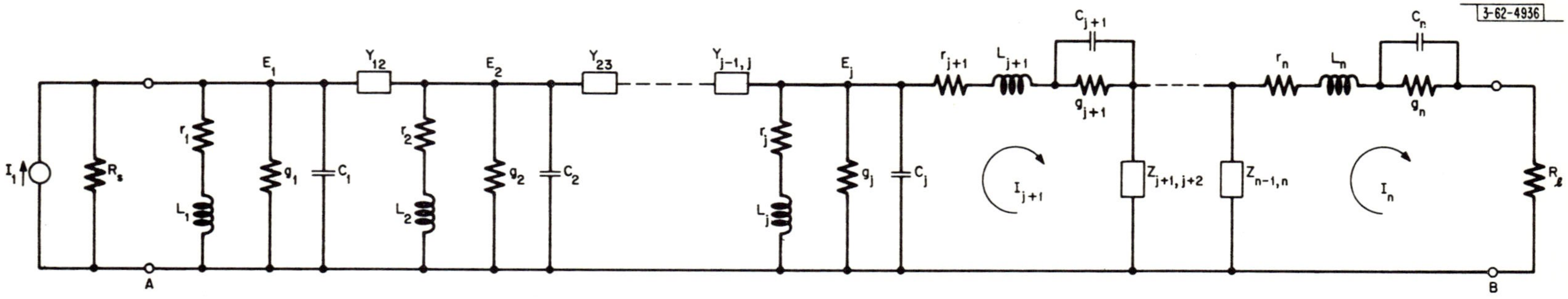


(c) Reactance-coupled parallel and series tuned circuits. With port 1 and loop j + 1 open, self-admittance of nodes 1 through j are resonant at $\omega_0 = 2\pi f_0$; with nodes j and datum and port n short circuited, self impedances of loops j + 1 through n are also resonant at $\omega_0$.

Fig. A-1. Practical realizations of narrow-band filters.

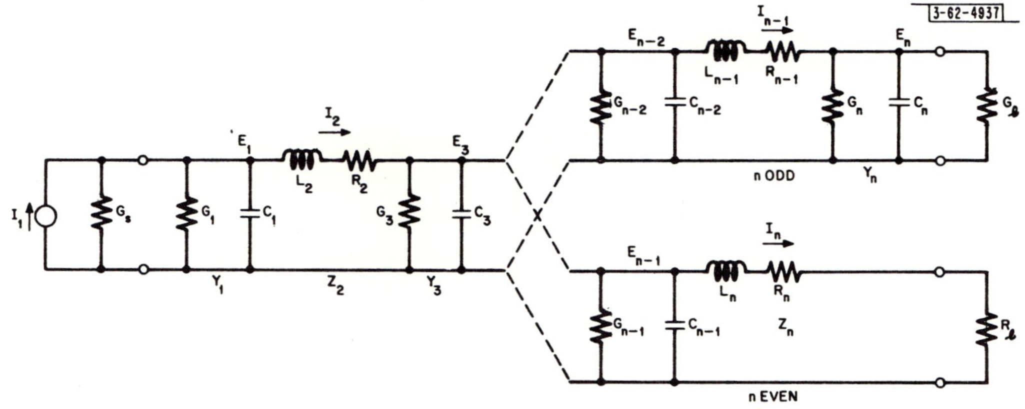


Fig. A-2. Circuit of lossy lowpass filter with no finite zeros of transmission.

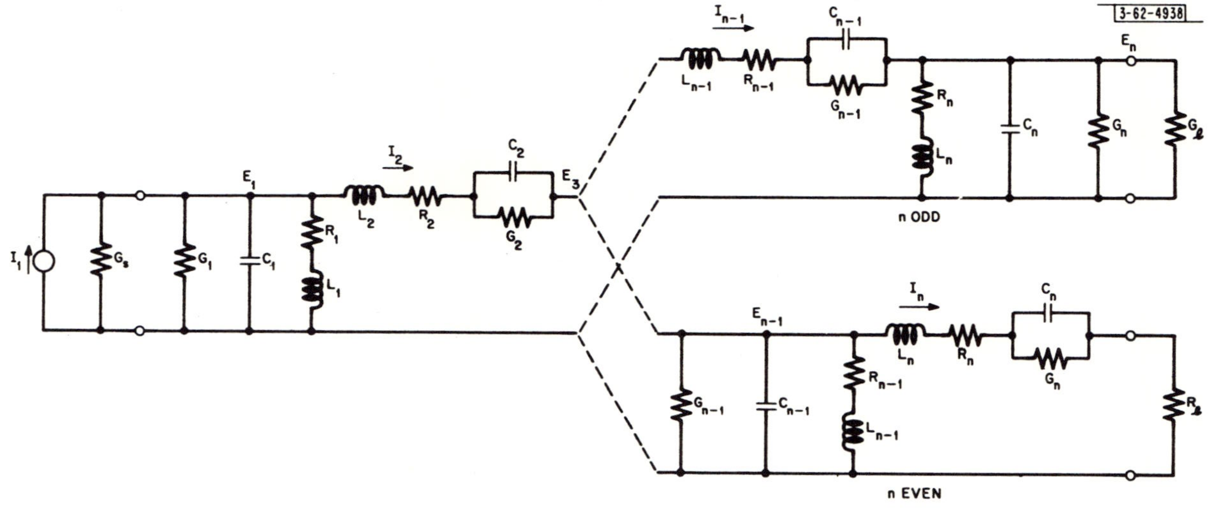


Fig. A-3. Lossy bandpass filter derived from lossy lowpass filter of Fig. A-2.

For the alignment of the bandpass circuit of Fig. A-3, the resonant circuits must be tuned individually to the center frequency. This can be accomplished by a procedure similar to that for narrow bandpass filters. Apply a sine wave of frequency $f_0$ from a high impedance source to the input port of the filter, put a high impedance voltmeter between node 1 and ground, and open link 2. Adjust $C_1$ or $L_1$ so that $E_1$ is maximum. Next close link 2 and short circuit node 3 to ground. Adjust $C_2$ or $L_2$ so that $E_1$ is minimum. Remove the short circuit from node 3 and open link 4. Adjust $C_3$ or $L_3$ so that $E_1$ is maximum. Continue until the end of the network is reached, and be sure that the load is disconnected when the last reactance is adjusted. The reactance that remains fixed during the tuning procedure should have a value that is within 3 or 5 percent of its theoretical value. An alternate procedure is to place each resonant circuit between a resistive sine wave generator of frequency $f_0$ and a resistive load. Then the variable reactance is adjusted for maximum load voltage with series resonant circuits and minimum load voltage with parallel resonant circuits. The tuning error should be less than $10^{-3} \times f_0$ in either case.

# APPENDIX B
# ANALYSIS OF DISSIPATIVE, DOUBLY TERMINATED BANDPASS AND LOWPASS LADDER FILTERS

In this appendix transfer functions of the narrow band networks of Fig. A-1 will be derived using standard matrix methods of network analysis along with narrow band approximations for the impedances of the tuned circuits. These transfer functions are valid in the vicinity of the pass band. A similar analysis is made of the lossy lowpass ladder networks of Fig. A-2 (with no finite zeros of transmission) to show the equivalence in form of the transfer functions of these two types of filters. Finally, a method of deriving a lossy wide bandpass filter from a uniformly lossy lowpass filter is presented. Thus, with the proper definition of the design parameters, the results presented in this report can be used for the design of filters of each type. The results are not new but a few new and interesting points are brought up along the way. In addition, the process of frequency normalization makes the design parameters arise naturally during the course of analysis.

## I. NARROW BANDPASS FILTERS

For the circuit of Fig. A-1(a), let $\mathfrak{I}$ be the node source-current vector (or column matrix), $\mathfrak{E}$ the node voltage vector (or column matrix), and $\mathfrak{Y}$ the node admittance matrix. Thus, the node voltage equations can be written compactly as

$$\mathfrak{I} = \mathfrak{Y}\mathfrak{E} \tag{B-1}$$

where

$$\mathfrak{I} = \begin{bmatrix} I_1 \\ I_2 \\ \cdot \\ \cdot \\ \cdot \\ I_n \end{bmatrix} = \begin{bmatrix} I_1 \\ 0 \\ \cdot \\ \cdot \\ \cdot \\ 0 \end{bmatrix} \tag{B-2}$$

$$\mathfrak{E} = \begin{bmatrix} E_1 \\ E_2 \\ \cdot \\ \cdot \\ \cdot \\ E_n \end{bmatrix} \tag{B-3}$$

and

$$
\mathcal{Y} = \begin{bmatrix}
Y_{11} & -Y_{12} & 0 & \cdots & \cdots & 0 \\
-Y_{12} & Y_{22} & -Y_{23} & 0 & \cdots & 0 \\
0 & -Y_{23} & Y_{33} & & & \vdots \\
\vdots & 0 & \ddots & \ddots & & 0 \\
\vdots & \vdots & & Y_{n-1,n-1} & & -Y_{n-1,n} \\
0 & 0 \cdots 0 & & -Y_{n-1,n} & & Y_{nn}
\end{bmatrix} . \qquad \text{(B-4)}
$$

$\mathcal{Y}$ is an n by n matrix whose only nonzero elements are on the main diagonal and on the two diagonals adjacent to it. This occurs because only the coupling between nonadjacent nodes is zero. Equation (B-1) can be solved formally for any of the node voltages $E_i$. By inverting it we obtain $\mathcal{E} = \mathcal{Y}^{-1}\mathcal{I}$, where $\mathcal{Y}^{-1}$ is the inverse of $\mathcal{Y}$. So for $E_i$ we obtain,

$$
E_i = I_1 \mathcal{Y}_{1i} / |\mathcal{Y}| \qquad \text{(B-5)}
$$

where $|\mathcal{Y}|$ is the determinant of $\mathcal{Y}$, and $\mathcal{Y}_{1i}$ is the cofactor of the element in the first row and $i^{th}$ column of $|\mathcal{Y}|$. For i = n we have $E_n = I_1 \mathcal{Y}_{1n} / |\mathcal{Y}|$ where

$$
\mathcal{Y}_{1n} = (-1)^{n+1} \begin{bmatrix}
-Y_{12} & Y_{22} & -Y_{23} & 0 & \cdots & 0 \\
0 & -Y_{23} & Y_{33} & & & \vdots \\
\vdots & 0 & -Y_{34} & & & \vdots \\
\vdots & \vdots & 0 & \ddots & & 0 \\
\vdots & \vdots & \vdots & \ddots & & -Y_{n-2,n-1} \\
\vdots & \vdots & \vdots & & -Y_{n-2,n-1} & Y_{n-1,n-1} \\
0 & 0 & 0 & \cdots & 0 & -Y_{n-1,n}
\end{bmatrix} \qquad \text{(B-6)}
$$

Evaluation of this (triangular) determinant is easily carried out by expanding it by elements of the first column and continuing likewise for the lower order determinants. The result is the product of the main diagonal elements. Therefore,

$$\mathcal{Y}_{1n} = (-1)^{n+1}(-Y_{12})\ldots(-Y_{n-1,n}) = Y_{12}Y_{23}\cdots Y_{n-1,n} \quad . \tag{B-7}$$

For this narrow band situation it is assumed that the reactive coupling admittances are constant and have a value equal to that at the center frequency. Let

$$Y_{i,i+1} = iB_{i,i+1} \quad . \tag{B-8}$$

Thus, if the coupling is capacitive, $B_{i,i+1} = \omega_0 C_{i,i+1}$, and if it is inductive, $B_{i,i+1} = -(\omega_0 L_{i,i+1})^{-1}$. So with Eq. (B-8) $\mathcal{Y}_{1n}$ is

$$\mathcal{Y}_{1n} = i^{n-1}B_{12}B_{23}\ldots B_{n-1,n} \quad . \tag{B-9}$$

Since this is as far as it is necessary to carry $\mathcal{Y}_{1n}$, we now consider $|\mathcal{Y}|$. The narrow band approximation will be applied to the main diagonal elements of $|\mathcal{Y}|$, and $|\mathcal{Y}|$ itself will be reduced to a normal form.

A main diagonal element of $|\mathcal{Y}|$ is $Y_{ii}$, and the narrow band approximation will be made by expanding $Y_{ii}(s)$ in a Taylor series about $s_0 = i\omega_0$,

$$Y_{ii}(s) = Y_{ii}(s_0) + (s - s_0)\left(\frac{dY_{ii}}{ds}\right)_{s=s_0} + \ldots \quad ,$$

and retaining only these two terms. For $Y_{ii}(s_0)$ we obtain,

$$Y_{ii}(s_0) = \omega_0 C_{ii} D_i \left(1 - i\frac{r_i}{\omega_0 L_{ii}}\right)^{-1} \approx \omega_0 C_{ii} D_i \quad ,$$

and for the derivative we obtain,

$$\left(\frac{dY_{ii}}{ds}\right)_{s=s_0} = 2C_{ii}\left(1 + i\frac{g_i}{\omega_0 C_{ii}} + \ldots\right) \approx 2C_{ii} \quad .$$

So for $Y_{ii}(s)$ we have approximately,

$$Y_{ii}(s) = C_{ii}\,[\omega_0 D_i + 2(s - i\omega_0)] \quad . \tag{B-10}$$

At this point it is convenient to introduce the normalized frequency variable $\lambda = s - i\omega_0/(w/2)$, where $w$ (rad/sec) is the desired bandwidth of the filter. For example, for a Butterworth filter $w$ might be its 3 dB-bandwidth, or for a Chebyshev filter $w$ might be its ripple-bandwidth. The choice of $w$ and its relation to the design data is explained in Sec. II. So $\lambda$ is the deviation from the center frequency in units of half the filter bandwidth. The variation of $Y_{ii}$ with $\lambda$ will be designated $y_{ii} = y_{ii}(\lambda)$, so

$$y_{ii} = C_{ii}(\omega_0 D_i + w\lambda) = wC_{ii}(\lambda + d_i) \quad . \tag{B-11}$$

Since this is the desired form for the main diagonal elements of $|\mathcal{Y}|$, now we will look at $|\mathcal{Y}|$ itself.

By using Eqs. (B-8) and (B-11) in the matrix $\mathcal{Y}$ (Eq. B-4) $|\mathcal{Y}|$ now is

$$|\mathcal{Y}| = \begin{vmatrix} wC_{11}(\lambda + d_1) & -iB_{12} & 0 & \cdots & \cdots & 0 \\ -iB_{12} & wC_{22}(\lambda + d_2) & -iB_{23} & 0 & \cdots & 0 \\ 0 & -iB_{23} & wC_{33}(\lambda + d_3) & & & \vdots \\ \vdots & 0 & \ddots & & & \vdots \\ \vdots & \vdots & & \ddots & & 0 \\ \vdots & \vdots & & & wC_{n-1,n-1}(\lambda + d_{n-1}) & -iB_{n-1,n} \\ 0 & 0 & \cdots & 0 & -iB_{n-1,n} & wC_{nn}(\lambda + d_n) \end{vmatrix} . \tag{B-12}$$

There is only one more step before we are finished working on $|\mathcal{Y}|$, and that is to factor the quantity $\sqrt{wC_{ii}}$ from the $i^{th}$ row and the $i^{th}$ column of $|\mathcal{Y}|$. The main diagonal elements then become $\lambda + d_i$, and the other elements become:

$$-i\frac{B_{12}}{w\sqrt{C_{11}C_{22}}} = -ik_{12} \quad , \quad -i\frac{B_{23}}{w\sqrt{C_{22}C_{33}}} = -ik_{23} \quad , \quad \text{etc.}$$

The determinant $|\mathcal{Y}|$ then is

$$|\mathcal{Y}| = w^n C_{11}C_{22}\cdots C_{nn} \begin{vmatrix} (\lambda + d_1) & -ik_{12} & 0 & \cdots & \cdots & 0 \\ -ik_{12} & (\lambda + d_2) & -ik_{23} & 0 & \cdots & 0 \\ 0 & -ik_{23} & (\lambda + d_3) & & & \vdots \\ \vdots & 0 & \ddots & & & \vdots \\ \vdots & \vdots & & \ddots & & 0 \\ \vdots & \vdots & & & (\lambda + d_{n-1}) & -ik_{n-1,n} \\ 0 & 0 & \cdots & 0 & -ik_{n-1,n} & (\lambda + d_n) \end{vmatrix} \tag{B-13}$$

The determinant on the right side of this equation is an $n^{th}$ order polynomial in $\lambda$, $Q_n(\lambda) = \lambda^n + q_{n-1}\lambda^{n-1} + \ldots + q_1\lambda + q_0$, whose coefficients are real, and it will be called the characteristic polynomial. It is not obvious that the coefficients are real, so we will make a small digression to prove this. At the same time a result is obtained that will be useful later to show the equivalence of the transfer functions of the filters of Fig. A-1.

The determinant in Eq. (B-13) is of the form

$$
\Delta^{(n)} = \begin{vmatrix}
a_{11} & a_{12} & 0 & \cdots & \cdots & 0 \\
a_{21} & a_{22} & a_{23} & 0 & \cdots & 0 \\
0 & a_{32} & a_{33} & & & \vdots \\
\vdots & 0 & \ddots & & & \vdots \\
\vdots & \vdots & & \ddots & & 0 \\
\vdots & \vdots & & & a_{n-1,\,n-1} & a_{n-1,\,n} \\
0 & 0 & \cdots & 0 & a_{n,\,n-1} & a_{nn}
\end{vmatrix}
$$

where the superscript n is used to denote that it has n rows and n columns. First, expand $\Delta^{(n)}$ in terms of the elements of the $n^{th}$ column to obtain

$$
\Delta^{(n)} = a_{nn}A_{nn} + a_{n-1,\,n}A_{n-1,\,n} \quad .
$$

The minor $A_{nn}$ is just the determinant $\Delta^{(n-1)}$, and by crossing out the $(n-1)^{st}$ row and $n^{th}$ column in $\Delta^{(n)}$ we see that the minor $A_{n-1,\,n} = -a_{n,\,n-1}\Delta^{(n-2)}$. Thus we obtain the result desired:

$$
\Delta^{(n)} = a_{nn}\Delta^{(n-1)} - a_{n,\,n-1}a_{n-1,\,n}\Delta^{(n-2)} \quad . \tag{B-14}
$$

By applying this result to $Q_n(\lambda)$ we obtain

$$
Q_n(\lambda) = (\lambda + d_n)\, Q_{n-1}(\lambda) + k^2_{n-1,\,n}Q_{n-2}(\lambda) \quad . \tag{B-15}
$$

A continuation of this process easily shows that $Q_n(\lambda)$ does have real coefficients. In addition, by starting the expansion of $\Delta^{(n)}$ with the elements of the $n^{th}$ column we have obtained in Eq. (B-15) a recursion formula for the polynomial $Q_n(\lambda)$. We can now return to the main stream of thought. Equation (B-13) is

$$
|\mathcal{Y}| = w^n C_{11}C_{22} \cdots C_{nn}Q_n(\lambda) \quad . \tag{B-16}
$$

With Eqs. (B-5), (B-9), and (B-16) we have for the transfer function

$$
\frac{E_n}{I_1} = \frac{i^{n-1}B_{12}B_{23} \cdots B_{n-1,\,n}}{w^n C_{11}C_{22} \cdots C_{nn}Q_n(\lambda)} = \frac{i^{n-1}k_{12}k_{23} \cdots k_{n-1,\,n}}{w\sqrt{C_{11}C_{nn}}\,Q_n(\lambda)} \quad , \tag{B-17}
$$

which is the desired form. Since the coefficients of the polynomial $Q_n(\lambda)$ are functions of the normalized coupling coefficients and dissipation factors, these $2n - 1$ parameters determine the frequency behavior of the transfer function. In Eq. (B-17) the frequency scale is normalized, but the multiplying factor is not. For example, the actual ratio between output voltage and input current at the center frequency ($s = i\omega_0$ or $\lambda = 0$) is

$$
\left.\frac{E_n}{I_1}\right|_{\lambda=0} = \frac{i^{n-1}k_{12}k_{23} \cdots k_{n-1,\,n}}{w\sqrt{C_{11}C_{nn}}\,q_0} = \frac{i^{n-1}\Gamma}{w\sqrt{C_{11}C_{nn}}} \quad . \tag{B-18}
$$

Since $q_0$ and the $k_{i,\,i+1}$'s depend only on the normalized frequency behavior of the filter and not on the impedance level of the circuit, the factor $(k_{12}k_{23} \cdots k_{n-1,\,n}/q_0)$ in Eq. (B-18) is defined as $\Gamma$, a normalized quantity that will be convenient to have available for determining the filter gain; $\Gamma$ is given with each set of data. The gain of the filter at the center frequency then is

$$\left| \frac{E_n}{I_1} \right|_{\lambda=0} = \frac{\Gamma}{w\sqrt{C_{11}C_{nn}}} \quad . \tag{B-19}$$

We next consider the filter of Fig. A-1(b). Since the circuits of Figs. A-1(a) and A-1(b) are duals, we can obtain any result necessary by inserting the dual quantities into the equations just derived. The normalized parameters change only from the point of view of their definitions in terms of the circuit element values. The result of major interest is the transfer function $I_n/E_1$ which, from Eq. (B-17), is

$$\frac{I_n}{E_1} = \frac{i^{n-1}k_{12}k_{23} \cdots k_{n-1,\,n}}{w\sqrt{L_{11}L_{nn}}Q_n(\lambda)} = \frac{i^{n-1}q_0\Gamma}{w\sqrt{L_{11}L_{nn}}Q_n(\lambda)} \tag{B-20}$$

and that the quantity (B-19) is also of interest here in determining the filter gain. Since the concept of duality is widely known, it is unnecessary to make any further remarks about the analysis of this circuit.

The circuit in Fig. A-1(c) is a combination of the two circuits that were just considered. It is useful when one has a high resistance source and a low resistance load or, by turning it around, a low resistance source and a high resistance load. The analysis of this circuit will be carried out using as the unknowns the node voltages $E_1, \ldots E_i$ and the loop currents $I_{i+1}, \ldots I_n$, with node current equations for the left-hand part of the circuit and loop voltage equations for the right hand part. For node $i$ and loop $i+1$ the following equations involve both $E_i$ and $I_{i+1}$.

$$\begin{aligned} &\text{node } i, && 0 = -Y_{i-1,\,i}E_{i-1} + Y_{ii}E_i + I_{i+1} \\ &\text{loop } i+1, && 0 = -E_i + Z_{i+1,\,i+1}I_{i+1} - Z_{i+1,\,i+2}I_{i+2} \quad . \end{aligned} \tag{B-21}$$

In terms of the source vector

$$\mathcal{Q} = \begin{bmatrix} I_1 \\ 0 \\ \cdot \\ \cdot \\ \cdot \\ 0 \end{bmatrix}$$

which has $n$ components, the node voltage and loop current vector

$$\mathcal{P} = \begin{bmatrix} E_1 \\ \cdot \\ \cdot \\ \cdot \\ E_i \\ I_{i+1} \\ \cdot \\ \cdot \\ \cdot \\ I_n \end{bmatrix}$$

and the n by n circuit matrix $\mathfrak{M}$, the matrix equation for the circuit is

$$\mathcal{Q} = \mathfrak{M}\mathcal{P} \quad .$$

The matrix $\mathfrak{M}$ is given in Eq. (B-22).

$$\mathfrak{M} = \begin{bmatrix}
wC_{11}(\lambda + d_1) & -iB_{12} & 0 & \cdots & \cdots & \cdots & \cdots & \cdots & 0 \\
-iB_{12} & wC_{22}(\lambda + d_2) & -iB_{23} & 0 & \cdots & \cdots & \cdots & \cdots & 0 \\
0 & -iB_{23} & \ddots & & & & & & \vdots \\
\vdots & 0 & & \ddots & 0 & & & & \vdots \\
\vdots & \vdots & & & -iB_{i-1,\, i} & 0 & & & \vdots \\
\vdots & \vdots & & -iB_{i-1,\, i} & wC_{ii}(\lambda + d_i) & 1 & 0 & & \vdots \\
\vdots & \vdots & & 0 & -1 & wL_{i+1,\, i+1}(\lambda + d_{i+1}) & -iX_{n+1,\, n+2} & & \vdots \\
\vdots & \vdots & & & 0 & -iX_{n+1,\, n+2} & \ddots & & 0 \\
\vdots & \vdots & & & & & \ddots & wL_{n-1,\, n-1}(\lambda + d_{n-1}) & -iX_{n-1,\, n} \\
0 & 0 & \cdots & \cdots & \cdots & \cdots & 0 & -iX_{n-1,\, n} & wL_{nn}(\lambda + d_n)
\end{bmatrix} \tag{B-22}$$

The narrow band approximation has been applied to the elements of $\mathfrak{M}$. The formal solution for $I_n$ is

$$I_n = I_1 \mathfrak{M}_{1n} / |\mathfrak{M}| \tag{B-23}$$

where $|\mathfrak{M}|$ is the determinant of $\mathfrak{M}$, and $\mathfrak{M}_{1n}$ is the cofactor of the element in the first row and $n^{th}$ column of $|\mathfrak{M}|$. We see that $\mathfrak{M}_{1n}$ is

$$\mathfrak{M}_{1n} = (-1)^{n+1}(-iB_{12})\ldots(-iB_{i-1,\,i})(-1)(-iX_{i+1,\,i+2})\ldots(-iX_{n-1,\,n})$$

$$= -i^n B_{12}\ldots B_{i-1,\,i}X_{i+1,\,i+2}\ldots X_{n-1,\,n} \quad . \tag{B-24}$$

This completes the reduction of the cofactor $\mathfrak{M}_{1n}$, and now as before we will factor the rows and columns of $|\mathfrak{M}|$ so that the main diagonal has elements of the form $\lambda + d_i$. From Eq. (B-22) we see that this is done by factoring $\sqrt{wC_{jj}}$ from the $j^{th}$ row and the $j^{th}$ column of $|\mathfrak{M}|$ for $j = 1, 2 \ldots i$, and by factoring $\sqrt{wL_{kk}}$ from the $k^{th}$ row and the $k^{th}$ column of $|\mathfrak{M}|$ for $k = i + 1, \ldots n$. This will automatically define for us the normalized coupling coefficient between the $i^{th}$ node and the $(i+1)^{st}$ loop as the element in the $i^{th}$ row and $(i+1)^{st}$ column of this modified determinant. The result of carrying out this process is given in Eq. (B-25).

$$|\mathfrak{M}| = w^n C_{11}\ldots C_{ii}L_{i+1,\,i+1}\ldots L_{nn}\begin{vmatrix} (\lambda+d_1) & -ik_{12} & 0 & \cdots & \cdots & \cdots & 0 \\ -ik_{12} & (\lambda+d_2) & \ddots & & & & \vdots \\ 0 & \ddots & \ddots & & & & \vdots \\ \vdots & & \ddots\;(\lambda+d_i) & \dfrac{(C_{ii}L_{i+1,\,i+1})^{-1/2}}{w} & & & \vdots \\ \vdots & & -\dfrac{(C_{ii}L_{i+1,\,i+1})^{-1/2}}{w} & (\lambda+d_{i+1}) & & & \vdots \\ \vdots & & & & \ddots & & 0 \\ \vdots & & & & \cdot\;(\lambda+d_{n-1}) & & -ik_{n-1,\,n} \\ 0 & \cdots & \cdots & \cdots & 0 & -ik_{n-1,\,n} & (\lambda+d_n) \end{vmatrix} \tag{B-25}$$

So we see that

$$k_{i,\,i+1} = \frac{1}{w(C_{ii}L_{i+1,\,i+1})^{1/2}} \quad . \tag{B-26}$$

The determinant in Eq. (B-25) is not quite like that in Eq. (B-13), but since in its expansion [see Eq. (B-14)] the elements off the main diagonal occur only as products $(-a_{k,\,k-1}a_{k-1,\,k})$ the determinant in Eq. (B-25) is the same polynomial, $Q_n(\lambda)$, as that in Eq. (B-13). Thus

$$|\mathfrak{M}| = w^n C_{11}\ldots C_{ii}L_{i+1,\,i+1}\ldots L_{nn}Q_n(\lambda) \quad . \tag{B-27}$$

So for the transfer function $I_n/I_1$ we obtain from Eqs. (B-23), (B-24) and (B-27),

$$\frac{I_n}{I_1} = \frac{-i^n B_{12}\ldots B_{i-1,\,i}X_{i+1,\,i+2}\ldots X_{n-1,\,n}}{w^n C_{11}\ldots C_{ii}L_{i+1,\,i+1}\ldots L_{nn}Q_n(\lambda)}$$

$$= \frac{-i^n k_{12}k_{23}\ldots k_{n-1,\,n}}{w\sqrt{C_{11}L_{nn}}\,Q_n(\lambda)} = \frac{-i^n q_0 \Gamma}{w\sqrt{C_{11}L_{nn}}\,Q_n(\lambda)} \tag{B-28}$$

which is the same form as that of the other two filter circuits.

Now, if instead of the current source $I_1$ at node 1 in Fig. A-1(c), we have a voltage source, $E_n$, in loop n whose positive direction is counter to that of the positive direction of $I_n$, we have a circuit for use with a low resistance source and a high resistance load. The source vector $\mathcal{Q}$ is

$$\mathcal{Q} = \begin{bmatrix} 0 \\ \cdot \\ \cdot \\ \cdot \\ 0 \\ -E_n \end{bmatrix}$$

and the transfer function $E_1/E_n = -\mathfrak{M}_{n1}/|\mathfrak{M}|$. It is easy to show that $E_1/E_n$ is identical to the expression for $I_n/I_1$ in Eq. (B-28).

Thus, the transfer functions of the narrow bandpass filters in Fig. A-1 have been derived and are given by Eqs. (B-17), (B-20) and (B-28). By using the appropriate definitions for coupling coefficient and dissipation factor, all the transfer functions have the same form. In fact, they contain the common factor, $i^n(k_{12}k_{23}\ldots k_{n-1,n})/wQ_n(\lambda)$, so that all can be designed from the same data, which are presented in this report for a wide variety of often used transfer functions.

This completes the analysis of the narrow bandpass filters. Now we will make a brief analysis of lossy lowpass filters and then discuss a method for deriving a lossy wide bandpass filter from a lowpass filter.

## II. LOWPASS FILTERS

The lowpass circuit to be considered is shown in Fig. A-2. All reactive elements have dissipation as shown and the circuit has no finite transmission zeros. We will be concerned with finding the transfer function of this circuit and showing that it is similar in form to that of the narrow band filters. The first component at the input is assumed to be a capacitor and the output is as shown, depending on whether n is odd or even. The dual circuits are obtained easily and their parameters will be obvious so they will not be discussed. The choice of circuit variables might at first appear to be awkward, but once the matrix for the equilibrium equations is written, it becomes a simple matter to obtain the desired result.

The variables to be used, the node voltages and link currents, are shown in Fig. A-2. Either the node voltages or the link currents alone are sufficient to determine the network behavior, but by using both we can show easily that the transfer function of the lowpass network of Fig. A-2 is equivalent in form to the normalized transfer function of the narrow band filters of Fig. A-1. The equilibrium equations for the lowpass network are

$$I_1 = Y_1E_1 + I_2$$

$$0 = -E_1 + Z_2I_2 + E_3$$

$$0 = -I_2 + Y_3E_3 + I_4$$

$$\vdots$$

$$\left.\begin{aligned} 0 &= -E_{n-2} + Z_{n-1}I_{n-1} + E_n \\ 0 &= -I_{n-1} + Y_nE_n \end{aligned}\right\} \quad \text{for } n \text{ odd} \quad . \qquad \text{(B-29)}$$

If n is even the last two equations of the set are:

$$\left.\begin{aligned} 0 &= -I_{n-2} + Y_{n-1}E_{n-1} + I_n \\ 0 &= -E_{n-1} + Z_nI_n \end{aligned}\right\} \quad \text{for } n \text{ even} \quad .$$

The branch admittances and impedances are:

$$Y_1 = sC_1 + G_1 + G_s = wC_1\left(\frac{s}{w} + \frac{G_1 + G_s}{wC_1}\right) = wC_1(\lambda + d_1)$$

$$Z_2 = sL_2 + R_2 = wL_2\left(\frac{s}{w} + \frac{R_2}{wL_2}\right) = wL_2(\lambda + d_2)$$

$$\vdots$$

$$Y_n = sC_n + G_n + G_\ell = wC_n\left(\lambda + \frac{G_n + G_\ell}{wC_n}\right) = wC_n(\lambda + d_n) \quad , \quad \text{for } n \text{ odd}$$

$$Z_n = sL_n + R_n + R_\ell = wL_n\left(\lambda + \frac{R_n + R_\ell}{wL_n}\right) = wL_n(\lambda + d_n) \quad , \quad \text{for } n \text{ even} \quad . \qquad \text{(B-30)}$$

The normalized frequency is $\lambda = s/w$. In the final form of these expressions, frequency and dissipation factors have been normalized by the bandwidth of the lowpass filter, w(rad/sec). For the present it will be assumed that n is odd. Later it will be easy to see what the result is for n even. In terms of the source vector

$$\mathcal{Q} = \begin{bmatrix} I_1 \\ 0 \\ \cdot \\ \cdot \\ \cdot \\ 0 \end{bmatrix} ,$$

the node voltage and link current vector

$$\mathcal{P} = \begin{bmatrix} E_1 \\ I_2 \\ E_3 \\ \cdot \\ \cdot \\ \cdot \\ E_n \end{bmatrix}$$

and the n by n circuit matrix $\mathfrak{M}$, the matrix equation for the circuit is

$$\mathcal{Q} = \mathfrak{M}\mathcal{P} \quad .$$

The matrix $\mathfrak{M}$ is

$$\mathfrak{M} = \begin{bmatrix} Y_1 & 1 & 0 & \cdot & \cdot & \cdot & \cdot & \cdot & \cdot & \cdot & 0 \\ -1 & Z_2 & 1 & 0 & \cdot & \cdot & \cdot & \cdot & \cdot & \cdot & 0 \\ 0 & -1 & Y_3 & & & & & & & & \cdot \\ \cdot & 0 & & \cdot & & & & & & & \cdot \\ \cdot & \cdot & & & \cdot & & & & & & \cdot \\ \cdot & \cdot & & & & \cdot & & & & & 0 \\ \cdot & \cdot & & & & & & Z_{n-1} & & & 1 \\ 0 & 0 & \cdot & \cdot & \cdot & \cdot & \cdot & 0 & -1 & & Y_n \end{bmatrix} \quad . \qquad \text{(B-31)}$$

The main diagonal elements of $\mathfrak{M}$ are the branch admittances and impedances whereas the elements of the diagonal just above it are all +1 and those of the diagonal just below it are all $-1$.

The transfer function of the network is

$$E_n/I_1 = \mathfrak{M}_{1n}/|\mathfrak{M}| \quad .$$

For $\mathfrak{M}$ of Eq. (B-31) it is easy to see that the cofactor $\mathfrak{M}_{1n} = 1$ for all n. Equation (B-14) could be used to expand $|\mathfrak{M}|$, but we do not need to carry the work this far. By using Eq. (B-30) in Eq. (B-31) we obtain for $|\mathfrak{M}|$,

$$|\mathfrak{M}| = \begin{vmatrix} wC_1(\lambda + d_1) & 1 & 0 & \cdots & \cdots & 0 \\ -1 & wL_2(\lambda + d_2) & \ddots & & & \vdots \\ 0 & \ddots & \ddots & \ddots & & \vdots \\ \vdots & & \ddots & \ddots & \ddots & 0 \\ \vdots & & & \ddots & wL_{n-1}(\lambda + d_{n-1}) & 1 \\ 0 & \cdots & \cdots & 0 & -1 & wC_n(\lambda + d_n) \end{vmatrix} . \qquad \text{(B-32)}$$

Now we will factor $\sqrt{wC_1}$ from the first row and the first column, $\sqrt{wL_2}$ from the second row and second column, and so on. Then we obtain Eq. (B-33).

$$|\mathfrak{M}| = w^n C_1 L_2 \cdots C_n \begin{vmatrix} (\lambda + d_1) & (w\sqrt{C_1L_2})^{-1} & 0 & \cdots & \cdots & \cdots & 0 \\ -(w\sqrt{C_1L_2})^{-1} & (\lambda + d_2) & (w\sqrt{L_2C_3})^{-1} & 0 & \cdots & \cdots & 0 \\ 0 & -(w\sqrt{L_2C_3})^{-1} & (\lambda + d_3) & \ddots & & & \vdots \\ \vdots & 0 & & \ddots & \ddots & & 0 \\ \vdots & \vdots & & & \ddots & \ddots & \vdots \\ \vdots & \vdots & & & & (\lambda + d_{n-1}) & (w\sqrt{L_{n-1}C_n})^{-1} \\ 0 & \cdots & 0 & \cdots & 0 & -(w\sqrt{L_{n-1}C_n})^{-1} & (\lambda + d_n) \end{vmatrix} . \qquad \text{(B-33)}$$

For this lowpass filter define the coupling coefficients as

$$k_{12} = \frac{1}{w\sqrt{C_1L_2}} , \quad k_{23} = \frac{1}{w\sqrt{L_2C_3}} , \quad \ldots , \quad k_{n-1,n} = \frac{1}{w\sqrt{L_{n-1}C_n}} . \qquad \text{(B-34)}$$

By using these definitions in Eq. (B-33) we obtain Eq. (B-35) for $|\mathfrak{M}|$.

$$|\mathfrak{M}| = \frac{w\sqrt{C_1C_n}}{k_{12}k_{23}\cdots k_{n-1,n}} \begin{vmatrix} (\lambda + d_1) & k_{12} & 0 & \cdots & \cdots & \cdots & 0 \\ -k_{12} & (\lambda + d_2) & k_{23} & 0 & \cdots & \cdots & 0 \\ 0 & -k_{23} & (\lambda + d_3) & \ddots & & & \vdots \\ \vdots & 0 & & \ddots & \ddots & & 0 \\ \vdots & \vdots & & & \ddots & (\lambda + d_{n-1}) & k_{n-1,n} \\ 0 & 0 & \cdots & \cdots & 0 & -k_{n-1,n} & (\lambda + d_n) \end{vmatrix} . \qquad \text{(B-35)}$$

Now by using Eq. (B-14) we see that the determinant in Eq. (B-35) is the polynomial $Q_n(\lambda)$ which has already appeared in the analysis of the narrow bandpass filters. Thus the transfer function is

$$\frac{E_n}{I_1} = \frac{k_{12}k_{23}\cdots k_{n-1,n}}{w\sqrt{C_1C_n}Q_n(\lambda)} = \frac{q_0\Gamma}{w\sqrt{C_1C_n}Q_n(\lambda)} \tag{B-36}$$

for n odd. For n even we find, in an analogous manner, that the transfer function $I_n/I_1$ is

$$\frac{I_n}{I_1} = \frac{k_{12}k_{23}\cdots k_{n-1,n}}{w\sqrt{C_1L_n}Q_n(\lambda)} = \frac{q_0\Gamma}{w\sqrt{C_1L_n}Q_n(\lambda)} \quad . \tag{B-37}$$

Thus the transfer functions of the lowpass filters, Eqs. (B-36) and (B-37), and those of the narrow band filters, Eqs. (B-17), (B-20), and (B-28), are identical in form (the narrow band filters have additional constant phase shift, as represented by the powers of i, that does not occur in the lowpass filters). Thus any of the filters can be designed from a set of data that applies to one of them. For a particular circuit the appropriate definition of the coupling coefficients and dissipation factors must be used.

## III. WIDE BANDPASS FILTERS

A bandpass filter of arbitrary bandwidth can be obtained from a lowpass filter by means of the well-known lowpass-to-bandpass (reactance) transformation (Weinberg,[8] p. 540). This approach is theoretically exact, but when parasitic loss in all reactances must be accounted for, it becomes approximate and the degree of approximation cannot be defined simply and precisely for the general case. A uniformly lossy bandpass filter of arbitrary bandwidth can be designed exactly by predistorting the bandpass transfer function, but for the purpose of this report we want to derive a lossy bandpass filter from a lossy lowpass filter. However, we can gain some feeling for what the effect of loss (and its distribution between the L's and C's) is by comparing the exact realization (in which half the reactances are uniformly lossy and the remaining reactances are lossless) with the case in which all reactances have the same dissipation factor.

Let us begin by considering the exact realization obtained by transformation of a uniformly lossy lowpass filter of Fig. A-2. The transformation introduces a lossless inductor in parallel with each capacitor of the lowpass filter and a lossless capacitor in series with each inductor of the lowpass filter. Thus we obtain a circuit like that in Fig. A-3 but with $R_1 = R_3 = \ldots = 0$ and $G_2 = G_4 = \ldots = 0$. It produces exactly the predicted transfer function, but in many cases it cannot be realized because we do not have components (usually inductors) that are sufficiently lossless. So the question naturally arises as to how the transfer function is changed if we redistribute the loss so that all reactances have some loss associated with them. In general this question cannot be answered without a detailed analysis of the circuit, but we are looking for a simpler yet useful answer. In the special case of uniform dissipation we can see in somewhat general terms how the poles of the resulting transfer function differ from those of the exact one. To do this we will obtain the final uniformly dissipative circuit in a way that allows us to keep track of its poles.

Begin with the uniformly lossy (dissipation factor = a) lowpass circuit considered above and remove all dissipation from the reactances to obtain the lossless predistorted lowpass filter. Then transform this lossless network to a bandpass network by the transformation used above, and add dissipation (= a/2) to all reactances. Now we have exactly the same circuit as that obtained by splitting the loss in each tuned circuit of the exact bandpass realization equally between the inductor and capacitor. By tracing the poles of the exact realization and of this approximation, we can compare their final locations, determine when the approximation is reasonably good, and predict the effect on the amplitude characteristic. It is unnecessary to trace the locations of all the poles; doing so for a pair of conjugate complex poles of the lowpass filter is sufficient.

Figure B-1(a) shows a conjugate pair of poles of the uniformly dissipative lowpass filter (×) and their predistorted location (⊗) (poles of the lossless, predistorted lowpass filter). Figure B-1(b) shows how the poles of (a) are transformed by the lowpass-to-bandpass transformation. The ×'s are the poles of the exact realization, and the ⊗'s are the poles of the approximation before uniform dissipation is added. Note that $\Theta_2 < \Theta_1$ and as a increases $\Theta_2$ decreases until $\Theta_2 = 0$ for $a = \alpha_0$. Predistortion causes the distance $\overline{CD}$ to be slightly smaller than the distance $\overline{AB}$. Figure B-1(c) shows the final position of the poles of the exact realization (×) [which are the same as in Fig. B-1(b)] and of the uniformly dissipative approximation (⊗). The latter are a distance a/2 to the left of their location in (b). Since $\Theta_2$ decreases as a increases, we see that the approximation becomes worse at the same time. Also, if a is fixed, the error is greatest for the lowpass poles with the smallest value of $\alpha_o$ (which is also the maximum value $a_{max}$ that a can have).

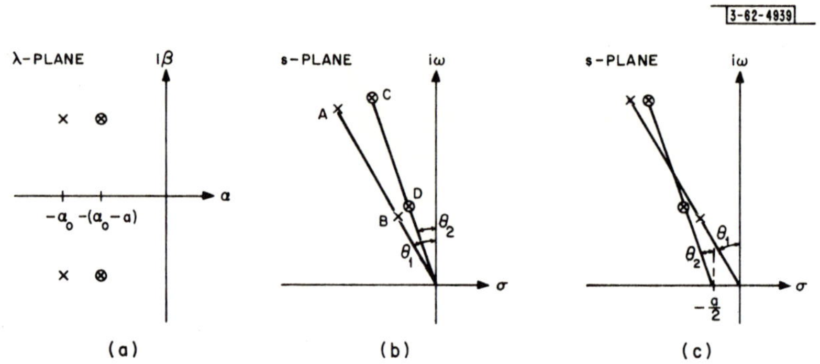


Fig. B-1. Pole locations for exact and approximate realization of dissipative bandpass filter. (a) Pair of conjugate complex poles (x) of uniformly dissipative lowpass filter and their predistorted location (⊗). (b) Third quadrant poles of (a) transformed to bandpass by reactance transformation. (c) Final location of poles of exact realization (x) and of uniformly dissipative approximation (⊗).

Thus far a constant ratio of bandwidth to center frequency has been implied, but as $w/\omega_0$ is increased and everything else is held constant the performance of the uniformly lossy filter improves. For very large values of $w/\omega_0$, both $\Theta_1$ and $\Theta_2$ are nearly 45°.

In the light of these observations we can see that if $a \ll a_{max}$ the approximation will be good (which agrees with intuition) and the distortion of the amplitude characteristic is an increase of the upper part of the pass band relative to the lower part. Since these errors are brought about by redistributing the loss in each tuned circuit as discussed above, it probably does not matter much how the losses in each tuned circuit are distributed as long as each tuned circuit has a dissipation factor equal to a which is $\ll a_{max}$. If a is less than 1/4 or 1/5 $a_{max}$ the resulting response will be satisfactory for most purposes.

## IV. SUMMARY

We have analyzed uniformly lossy narrow band and lowpass filters and have shown that their transfer functions can be put into identical forms if certain normalized parameters are defined appropriately. Also lossy bandpass filters of arbitrary bandwidth can be derived in a reasonably predictable way from the lowpass circuits if the reactances have small enough loss.

# APPENDIX C
## SOLUTION FOR THE DESIGN PARAMETERS ASSUMING UNIFORM DISSIPATION

In Appendix B we found the transfer functions for the circuits of Figs. A-1 and -2. By normalizing the frequency and appropriately defining the circuit design parameters, all transfer functions have identical dependence on these parameters (normalized coupling coefficients and dissipation factors). The polynominal $Q_n(\lambda)$, whose coefficients are functions of the k's and d's, determines the frequency characteristic of the filter. In order to design a filter, a set of coupling coefficients and dissipation factors must be found that will give the required coefficients of $Q_n(\lambda)$. This appendix gives the solutions to that design problem for n = 2, 3, 4, and 5. For n = 2, 3, and 4, Wagner[13] used a clever scheme to solve the design problem and his method is used here. For n = 5, Wagner's method fails, so a straightforward and general method originally proposed by Dishal[12] is used. In this approach coefficients of $Q_n(\lambda)$ are equated to the corresponding ones of a polynominal that produces the required transfer function. To the author's knowledge the solution presented here for lossy, doubly loaded five-pole filters has not been published before, but it seems unlikely that it has not been derived before.

In each case the solution is given in the following form: one or more of the normalized coupling coefficients is expressed as a function of the coefficient of $Q_n(\lambda)$ and the normalized dissipation factors of the circuit, and the remaining ones are expressed as functions of the preceding ones and the polynomial coefficients. This is a convenient form since, with the assumption of uniform dissipation, there are at most three dissipation factors involved in a design (d, a, and $\delta$) and the sum of these must be a constant equal to $q_{n-1}$. The choice of circuit components determines the level of uniform dissipation a, and so the sum of d plus $\delta$ is fixed. Thus, the coupling coefficients can be considered to be functions of the polynomial coefficients $q_i$ and the dissipation factor d with the dissipation factor a as a parameter. All coupling coefficients appear only in the squared form, so the following notation will be used.

$$k_{12}^2 = u \quad , \quad k_{23}^2 = v$$

$$k_{34}^2 = w \quad , \quad k_{45}^2 = x \quad .$$

Also, when uniform dissipation is imposed, the notation $d_1 = d$, $d_2 = \ldots = d_{n-1} = a$, and $d_n = \delta$ will be used. Since a is the unloaded dissipation factor, $a \leqslant d$ or $\delta$. We will assume that $d \leqslant \delta$ so we have, $a \leqslant d \leqslant \delta$. This assumption imposes no restriction on the results.

For $n \geqslant 3$, it is possible to have more than one set of u, v..., for the same values of d, a, and $\delta$, that will realize a given transfer function. An alternate viewpoint to these multiple solutions is provided by the insertion loss theory of uniformly lossy, reactive coupling networks. In this case multiple solutions arise because of the multiplicity of choices available for the zeros of the reflection coefficient at the input terminals of the network. Green[2] discusses these multiple solutions for lossless Butterworth and Chebyshev filters and finds ranges of d for which they are realizable. The results here verify these findings of Green, and multiple solutions for the lossy cases are also found.

### I. SOLUTION FOR n = 2

The desired polynomial is $Q_2(\lambda) = \lambda^2 + q_1\lambda + q_0$. Equating this to the one obtained from the circuit determinant yields

$$Q_2(\lambda) = (\lambda + d)(\lambda + \delta) + u = \lambda^2 + q_1\lambda + q_0 \quad . \tag{C-1}$$

Wagner's method is to set $\lambda = -d$ and $\lambda = -\delta$, thus we have

$$u = Q_2(-d) = Q_2(-\delta) \tag{C-2}$$

which has solutions for $d \neq \delta$. By equating coefficients of like powers of $\lambda$ in Eq. (C-1) and solving for u, a form recognizable as that of Eq. (C-2) is obtained but, especially for n = 3 and 4, Wagner's approach provides a more elegant solution. From the coefficients of $\lambda$ in Eq. (C-1) we have $d + \delta = q_1$, so the solution for n = 2 is

$$\delta = q_1 - d$$

$$u = Q_2(-d) \tag{C-3}$$

which is the only solution for n = 2.

## II. SOLUTION FOR n = 3

The polynomial obtained from the circuit determinant is $Q_3(\lambda) = (\lambda + d_1)(\lambda + d_2)(\lambda + d_3) + k_{12}^2(\lambda + d_3) + k_{23}^2(\lambda + d_1)$. By using the abbreviations mentioned above and equating this polynomial to the desired one, we obtain

$$Q_3(\lambda) = (\lambda + d)(\lambda + a)(\lambda + \delta) + u(\lambda + \delta) + v(\lambda + d)$$

$$= \lambda^3 + q_2\lambda^2 + q_1\lambda + q_0 \quad . \tag{C-4}$$

By equating the coefficients of $\lambda^2$ we obtain $q_2 = a + d + \delta$. By setting $\lambda = -d$ and then $\lambda = -\delta$ we obtain $Q_3(-d) = u(\delta - d)$ and $Q_3(-\delta) = v(d - \delta)$. So the solutions for u and v are

$$u = \frac{Q_3(-d)}{\delta - d}$$

$$v = -\frac{Q_3(-\delta)}{\delta - d} \tag{C-5}$$

or

$$u + v = q_1 - d\delta - a(d + \delta)$$

where

$$\delta = q_2 - (a + d) \quad \text{or} \quad \delta - d = q_2 - (a + 2d) \quad .$$

For the case $d = \delta$, Eq. (C-5) cannot be used easily as it is. Go back to Eq. (C-4) and write the lower line in factored form [$Q_3(\lambda)$ has a real root at $\lambda = -\sigma$ and a pair of conjugate roots at $\lambda = -\mu \pm i\sqrt{\nu^2 - \mu^2}$].

$$(\lambda + d)^2(\lambda + a) + (\lambda + d)(u + v) = (\lambda + \sigma)(\lambda^2 + 2\mu\lambda + \nu^2) \quad . \tag{C-6}$$

Since $(\lambda + d)$ is a factor of the left side and $(\lambda + \sigma)$ is a factor of the right side, d must equal $\sigma$ to satisfy the equation. By cancelling this factor in Eq. (C-6) and equating the remaining coefficients we have

$$a = 2\mu - \sigma$$

$$u + v = \nu^2 - a\sigma = \nu^2 + \sigma^2 - 2\mu\sigma \qquad \text{(C-7)}$$

with

$$d = \delta = \sigma \quad .$$

Equation (C-7) is the solution for $d = \delta$ and there are some interesting points about it. Note that only the sum $u + v$ is specified so that there is a continuum of solutions for u and v with fixed $d = \delta = \sigma$. Also for the two transfer functions of most interest, Butterworth and Chebyshev, $\sigma$ and $\mu$ are related by $\sigma = 2\mu$ so that by Eq. (C-7) the only solution is $a = 0$ and $u + v = \nu^2$. Feldtkeller,[1] Green,[2] and Wagner[13] mention this restricted solution in their investigations of Butterworth and Chebyshev filters, but their work, being restricted to these cases, does not show that solutions for $a > 0$ are possible for other transfer functions, i.e., cases where $2\mu > \sigma$. In particular, the maximally flat delay transfer function and the closely related Gaussian approximation of Dishal[5] are of great current interest and have $2\mu > \sigma$. On the other hand, for $2\mu < \sigma$ there is no realizable solution ($a < 0$), and this is a situation that one gets into when the three-pole Butterworth or Chebyshev filter is predistorted (with $d = \delta$).

## III. SOLUTION FOR n = 4

The characteristic polynomial for n = 4 is

$$Q_4(\lambda) = (\lambda + d_1)(\lambda + d_2)(\lambda + d_3)(\lambda + d_4) + k_{12}^2(\lambda + d_3)(\lambda + d_4)$$
$$+ k_{23}^2(\lambda + d_1)(\lambda + d_4) + k_{34}^2(\lambda + d_1)(\lambda + d_2) + k_{12}^2 k_{34}^2 \quad .$$

By using the abbreviations and equating this to the desired polynomial we obtain

$$Q_4(\lambda) = \lambda^4 + q_3\lambda^3 + q_2\lambda^2 + q_1\lambda + q_0$$
$$= (\lambda + d)(\lambda + a)^2(\lambda + \delta) + u(\lambda + a)(\lambda + \delta) + v(\lambda + d)(\lambda + \delta)$$
$$+ w(\lambda + d)(\lambda + a) + uw \quad . \qquad \text{(C-8)}$$

Equating the coefficients of $\lambda^3$ gives

$$d + 2a + \delta = q_3 \quad . \qquad \text{(C-9)}$$

By setting $\lambda = -d$, $\lambda = -a$, and then $\lambda = -\delta$, we obtain

$$Q_4(-d) = u(a - d)(\delta - d) + uw \qquad \text{(C-10)}$$

$$Q_4(-a) = v(d - a)(\delta - a) + uw \qquad \text{(C-11)}$$

$$Q_4(-\delta) = w(\delta - a)(\delta - d) + uw \qquad \text{(C-12)}$$

respectively. Solve Eq. (C-10) for u

$$u = \frac{Q_4(-d)}{w - (d - a)(\delta - d)} \qquad \text{(C-13)}$$

and put the result into Eq. (C-12) to obtain a quadratic equation for w,

$$w^2 + w\left[\frac{Q_4(-d) - Q_4(-\delta)}{(\delta - a)(\delta - d)} - (\delta - d)(d - a)\right] + \frac{d-a}{\delta - a} Q_4(-\delta) = 0 \quad . \tag{C-14}$$

For selected values of a and d, $\delta$ is determined from Eq. (C-9), then Eq. (C-14) can be solved for two values of w; u is obtained from Eq. (C-13), and v is obtained by subtracting Eq. (C-12) from (C-11) and solving for v,

$$v = \frac{Q_4(-a) - Q_4(-\delta)}{(d-a)(\delta - a)} + w\,\frac{\delta - d}{d - a} \quad . \tag{C-15}$$

For n = 4, Eqs. (C-9) and (C-13) through (C-15) constitute the solution to the design problem. For any given pair of positive values for a and d, there can be at most two realizable solutions; in some cases there will be only one, and in others there will be none.

## IV. SOLUTION FOR n = 5

The characteristic polynomial for n = 5 is

$$\begin{aligned} Q_5(\lambda) = {} & (\lambda + d_1)(\lambda + d_2)(\lambda + d_3)(\lambda + d_4)(\lambda + d_5) + k_{12}^2(\lambda + d_3)(\lambda + d_4)(\lambda + d_5) \\ & + k_{23}^2(\lambda + d_1)(\lambda + d_4)(\lambda + d_5) + k_{34}^2(\lambda + d_1)(\lambda + d_2)(\lambda + d_5) \\ & + k_{45}^2(\lambda + d_1)(\lambda + d_2)(\lambda + d_3) + k_{12}^2 k_{34}^2(\lambda + d_5) \\ & + k_{12}^2 k_{45}^2(\lambda + d_3) + k_{23}^2 k_{45}^2(\lambda + d_1) \quad . \end{aligned}$$

By imposing uniform dissipation ($d_2 = d_3 = d_4 = a$) and using the abbreviations, and then equating this to the desired polynomial we obtain

$$\begin{aligned} Q_5(\lambda) &= \lambda^5 + q_4\lambda^4 + q_3\lambda^3 + q_2\lambda^2 + q_1\lambda + q_0 \\ &= (\lambda + d)(\lambda + a)^3(\lambda + \delta) + u(\lambda + a)^2(\lambda + \delta) + (v + w)(\lambda + d) \\ &\quad \times (\lambda + a)(\lambda + \delta) + x(\lambda + d)(\lambda + a)^2 + uw(\lambda + \delta) \\ &\quad + ux(\lambda + a) + vx(\lambda + d) \quad . \end{aligned} \tag{C-16}$$

Wagner's scheme for n = 3 and 4 does not work for n = 5, so here we will equate coefficients of like powers of $\lambda$ in Eq. (C-16) and so obtain the following set of equations.

$$q_4 = d + 3a + \delta \tag{C-17}$$

$$q_3 = 3a(a + d + \delta) + d\delta + u + v + w + x \tag{C-18}$$

$$q_2 = a^3 + 3a^2(d + \delta) + 3ad\delta + u(2a + \delta) + (v + w)(a + d + \delta) + x(2a + d) \tag{C-19}$$

$$\begin{aligned} q_1 = {} & a^3(d + \delta) + 3a^2 d\delta + u(a^2 + 2a\delta) + (v + w)[a(d + \delta) + d\delta] \\ & + x(a^2 + 2ad) + uw + ux + vx \end{aligned} \tag{C-20}$$

$$q_0 = a^3 d\delta + a^2(\delta u + dx) + ad\delta(v + w) + \delta uw + aux + dvx \quad . \tag{C-21}$$

Equation (C-17) determines $\delta$ from a choice of a and d. Then Eqs. (C-18) through (C-21) are four equations in four unknowns (u, v, w, and x) which can easily be reduced to a pair of quadratic equations in x and w. In this pair of equations terms involving $w^2$ do not appear because of the assumed uniform dissipation. This then allows the pair of quadratic equations to be reduced to a cubic equation in x. Without the assumption of uniform dissipation, the resulting pair of equations reduces only to a quartic equation in w or x.

The solution of the above set of equations is straightforward but tedious, and proceeds as follows. Rewrite Eqs. (C-18) and (C-19) as

$$R = q_3 - 3a(d + a + \delta) - d\delta = u + v + w + x \tag{C-22}$$

$$S = q_2 - a^3 - 3a^2(d + \delta) - 3ad\delta = A_1 u + B_1 v + C_1 w + D_1 x \tag{C-23}$$

where

$$A_1 = 2a + \delta \quad , \quad B_1 = C_1 = a + d + \delta \quad , \quad D_1 = 2a + d \quad . \tag{C-24}$$

Solve the pair (C-22) and (C-23) for u and v to obtain

$$u = \frac{B_1 R - S + (C_1 - B_1)\, w + (D_1 - B_1)\, x}{B_1 - A_1} \tag{C-25}$$

and

$$v = \frac{S - A_1 R - (C_1 - A_1)\, w - (D_1 - A_1)\, x}{B_1 - A_1} \quad . \tag{C-26}$$

Now rewrite Eq. (C-20) as

$$\begin{aligned} T &= q_1 - a^3(d + \delta) - 3a^2 d\delta \\ &= A_2 u + B_2 v + C_2 w + D_2 x + uw + ux + vx \end{aligned} \tag{C-27}$$

where

$$\begin{aligned} A_2 &= a^2 + 2a\delta \\ B_2 &= C_2 = a(d + \delta) + d\delta \\ D_2 &= a^2 + 2ad \quad . \end{aligned} \tag{C-28}$$

By substituting for u and v from Eqs. (C-25) and (C-26) into Eq. (C-27) and combining terms, we obtain (after performing some algebra)

$$\begin{aligned} T = &-x^2 - wx\left(1 - \frac{D_1 - B_1}{B_1 - A_1}\right) + x\left[D_2 + R + \frac{A_2(D_1 - B_1) + B_2(A_1 - D_1)}{B_1 - A_1}\right] \\ &+ w\left[C_2 + \frac{B_1 R - S + A_2(C_1 - B_1) + B_2(A_1 - C_1)}{B_1 - A_1}\right] + \frac{C_1 - B_1}{B_1 - A_1}\, w^2 \\ &+ \frac{S(B_2 - A_2) + R(A_2 B_1 - A_1 B_2)}{B_1 - A_1} \end{aligned} \tag{C-29}$$

This is one quadratic equation in w and x. After deriving the second one both will be expressed in their final form.

To obtain the second quadratic equation, rewrite Eq. (C-21) as

$$P = q_o - a^3 d\delta$$

$$= A_3 u + B_3 v + C_3 w + D_3 x + \delta uw + aux + dvx \qquad \text{(C-30)}$$

where

$$A_3 = a^2 \delta$$

$$B_3 = C_3 = ad\delta$$

$$D_3 = a^2 d \quad . \qquad \text{(C-31)}$$

By again substituting for u and v from Eqs. (C-25) and (C-26) but now into Eq. (C-30) and collecting like terms we obtain

$$P = \frac{x^2}{B_1 - A_1}\,[a(D_1 - B_1) - d(D_1 - A_1)]$$

$$+ \frac{wx}{B_1 - A_1}\,[\delta(D_1 - B_1) - d(C_1 - A_1) + a(C_1 - B_1)]$$

$$+ x\left[D_3 + \frac{S(d - a) + R(aB_1 - dA_1) + A_3(D_1 - B_1) - B_3(D_1 - A_1)}{B_1 - A_1}\right]$$

$$+ w\left[C_3 + \frac{\delta(B_1 R - S) - B_3(C_1 - A_1)}{B_1 - A_1}\right] + \frac{\delta(C_1 - B_1)}{B_1 - A_1}\,w^2$$

$$+ \frac{1}{B_1 - A_1}\,[S(B_3 - A_3) + R(A_3 B_1 - A_1 B_3)] \quad . \qquad \text{(C-32)}$$

This is the second quadratic equation in w and x, but, in order to keep them general up to this point, we have not imposed the conditions in either that $B_1 = C_1$, $B_2 = C_2$, and $B_3 = C_3$. These conditions arise from the assumption of uniform dissipation ($d_2 = d_3 = d_4 = a$), but if only $d_2 = d_4$ they are true also. However, we are interested in the former case only.

By using the known quantities (A's, B's, C's, and D's) from Eqs. (C-23), (C-28), and (C-31) in the quadratic equations (C-29) and (C-32), the quadratics can be reduced to the following forms:

$$a_1 x^2 + c_1 wx + d_1 x + e_1 w + f_1 = 0 \qquad \text{(C-33)}$$

and

$$a_2 x^2 + c_2 wx + d_2 x + e_2 w + f_2 = 0 \quad . \qquad \text{(C-34)}$$

Here the constants are

$$a_1 = 1 \quad , \quad c_1 = 1 + \frac{\delta - a}{d - a}$$

$$d_1 = -[R + (\delta - a)(\delta - d)] \quad , \quad e_1 = \frac{S - R(a + d + \delta)}{d - a}$$

$$f_1 = T + R(\delta^2 + a^2 + a\delta) - S(\delta + a)$$

$$a_2 = \delta - d - a \quad , \quad c_2 = -[d + \frac{\delta(\delta - a)}{d - a}]$$

$$d_2 = S + a(\delta - d)(\delta - a) - R(\delta + a) \quad , \quad e_2 = \frac{\delta}{d - a}[R(a + d + \delta) - S]$$

$$f_2 = a\delta [S - R(\delta + a)] - P \quad . \qquad \text{(C-35)}$$

Finally, a cubic equation for $x$ is obtained by solving Eq. (C-33) for $w$ and substituting the result into Eq. (C-34). The result of this is

$$w = -\frac{a_1x^2 + d_1x + f_1}{c_1x + e_1} \qquad \text{(C-36)}$$

and

$$(a_2c_1 - a_1c_2)\,x^3 + (c_1d_2 - c_2d_1 + e_1a_2 - e_2a_1)\,x^2$$
$$+(c_1f_2 - c_2f_1 + e_1d_2 - e_2d_1)\,x + (e_1f_2 - e_2f_1) = 0 \quad . \qquad \text{(C-37)}$$

The coefficients in Eq. (C-37) are functions of $d$, $a$ ($\delta = q_1 - d - a$), and the coefficients of $Q_n(\lambda)$, so after choosing values of these parameters, three values of $x$ are obtained from the solution of Eq. (C-37). For each of these values of $x$, $w$ is obtained from Eq. (C-36), and $u$ and $v$ are obtained from

$$u = \frac{(a + d + \delta)\,R - S - x(\delta - a)}{d - a} \qquad \text{(C-38)}$$

and

$$v = \frac{S - (2a + \delta)\,R + x(\delta - d)}{d - a} - w \qquad \text{(C-39)}$$

which are from Eqs. (C-25) and (C-26), respectively. Thus we have the solution for $n = 5$.

## V. MULTIPLE SOLUTIONS

We have seen above that there is the possibility of more than one solution for the coupling coefficients (with fixed $d$, $a$, and $\delta$) when $n = 3$, 4, or 5. For $n = 3$, one of the solutions is singular in the sense that only the sum of $u$ and $v$ is dependent on the dissipation factors. For $n = 4$ there can be two solutions, and for $n = 5$ there can be three solutions. This is the case for uniform dissipation.

If the dissipation is arbitrarily distributed, then for $n = 4$ there still can be no more than two solutions (see Wagner[13]). But for $n = 5$ there can now be four solutions. A term involving $w^2$ now appears in Eqs. (C-33) and (C-34), and the pair of simultaneous quadratic equations reduces to a quartic equation in $x$ or $w$. Since only positive values of $u$, $v$, $w$, and $x$ lead to

3-62-4940

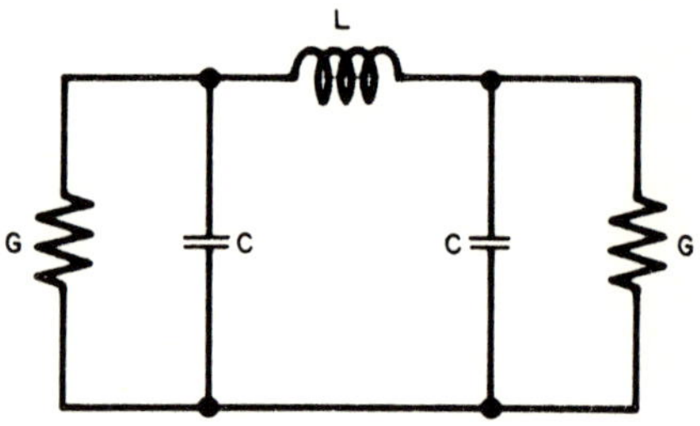


$$d = \delta = \frac{G}{wC}$$

$$u = v = \frac{1}{w^2 LC}$$

(a) Symmetric network for n = 3.

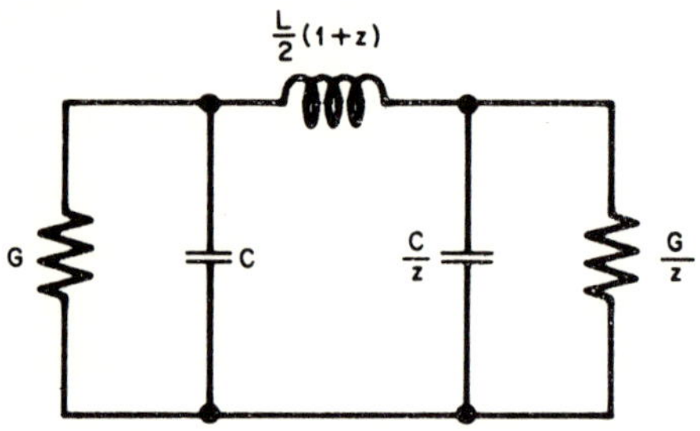


$$d' = \delta' = \frac{G}{wC}$$

$$v' = \frac{2z}{w^2 LC(1 + z)}$$

$$u' = \frac{2}{w^2 LC(1 + z)}$$

$$u' + v' = \frac{2}{w^2 LC}$$

(b) Transformed network for n = 3.

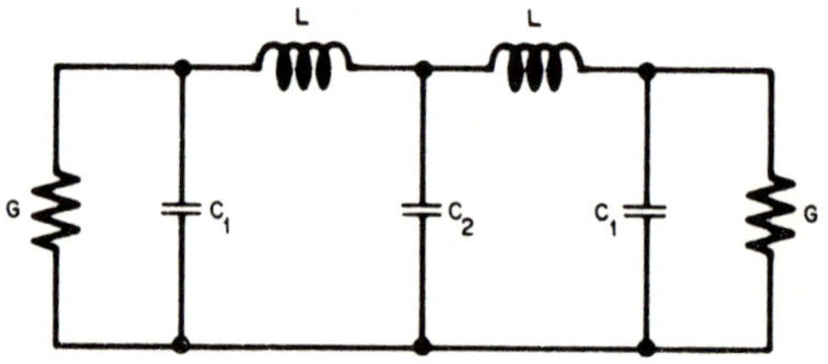


$$d = \delta = \frac{G}{wC_1}$$

$$u = x = \frac{1}{w^2 LC_1}$$

$$v = w = \frac{1}{w^2 LC_2}$$

(c) Symmetric network for n = 5.

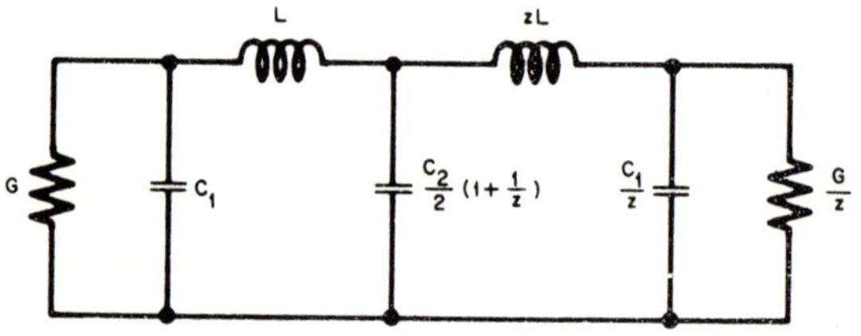


$$d' = \delta' = \frac{G}{wC_1}$$

$$w' = \frac{2}{w^2 LC_2(1 + z)}$$

$$u' = x' = \frac{1}{w^2 LC_1}$$

$$v' + w' = \frac{2}{w^2 LC_2}$$

$$v' = \frac{2z}{w^2 LC_2(1 + z)}$$

(d) Transformed network for n = 5.

Fig. C-1. Illustration of Takahasi-Weinberg transformation.

realizable networks, a set of solutions with a negative or complex value for any of these parameters is of no interest here.

The appearance of multiple solutions is an interesting aspect of this problem, and for lossless Butterworth and Chebyshev filters it has been investigated by Green.[2] For n = 3 and d = δ he did not observe the singular solution mentioned above. For n = 4 he obtained a second solution from a quadratic equation, as here, which for a BU 4 filter is realizable for the ratio δ/d between $\sqrt{2} - 1$ and $\sqrt{2} + 1$. Here we consider $\delta \geqslant d$ only; for $\delta < d$ the network is reversed. The BU 4 design curves show that the range for a realizable solution is in agreement with Green.

For n = 5 Green finds the second and third solutions by considering the problem from the point of view of modern network synthesis [see Weinberg,[7] Chaps. 2 and 12]. In this approach one starts with a transmission coefficient which is a rational function of frequency and is the ratio of load power to power available from the source. From this the reflection coefficient at the input of the terminated, lossless coupling network is determined, but it is not unique. The poles of the reflection coefficient are unique, but its zeros can be chosen in a number of ways. For each different and allowable choice of zeros, a different coupling network results, but all have the same transmission coefficient. Multiple solutions arise in this way from this approach.

For a five-pole Butterworth or Chebyshev filter, Green finds that there are four different realizable networks.* One of these is the normal solution for which all zeros of the reflection coefficient are in the left half λ-plane, and it is realizable for all positive values of δ/d. Two other solutions have certain ranges of realizability, and for a Butterworth filter these are $\sqrt{5} \geqslant \delta/d \geqslant 1/\sqrt{5}$ and $\sqrt{5} + 2 \geqslant \delta/d \geqslant \sqrt{5} - 2$. The results here agree with this. For the last solution Green finds that d = δ, u = x, and v = w, which is identical to the normal solution for d = δ. This is true but there is more to it. This is a particular case of a singular solution that is similar to the one for n = 3 above. For n = 5 the singular solution is δ = d, a = 0, u = x, and v + w = constant, and it can be shown that for Butterworth and Chebyshev filters, Eqs. (C-17) through (C-21) lead to this solution.

Takahasi[21] and Weinberg[22] have shown that if one takes the normal symmetrical circuit for n odd, bisects it, multiplies the impedance level of one half by a constant and then reconnects the two parts, the resulting network has a transfer coefficient that is unchanged except for a constant factor and a reflection coefficient at its input whose zeros, when taken in order of increasing magnitude of their real part, alternate between the right and left half planes. This is the arrangement of zeros that Green has in his fourth solution. Figure C-1 illustrates the effect of this transformation on the normal symmetrical circuits for n = 3 and 5. The unprimed parameters are those existing before the transformation occurs, and the primed parameters result after the network has been bisected, the impedance of the right half multiplied by z, and the two parts reconnected. Figures C-1(a) and (c) show the original networks, and Figs. C-1(b) and (d) show the transformed networks. For n = 3 we see that $d = \delta = d' = \delta'$ and $u + v = u' + v'$. For n = 5, $d = \delta = d' = \delta'$, $u = x = u' = x'$ and $v + w = v' + w'$. In general the design parameters have a symmetry that is unchanged by the transformation.

A similar singular solution occurs for any odd value of n. Thus, for lossless Butterworth and Chebyshev filters with n = 3 and 5, we see that the algebraic equations for the coupling

*Actually there are eight different ways to choose the zeros of reflection coefficient, but only four of these lead to different networks. The remaining four networks are the previous four turned end for end.

coefficients in Secs. C-II and -IV above lead to multiple solutions that correspond in a one-to-one manner with those obtained from modern network synthesis.

For Bessel filters an examination of the design curves shows that for each value of d only one solution is realizable (with the exception of BE 4 for $4.24 \leqslant d \leqslant 5.0$ and BE 5 for $6.8 \leqslant d \leqslant 7.5$ where there are two realizable solutions). Since there must obviously be several possible arrangements of zeros of the reflection coefficient, the network synthesis approach will yield a corresponding number of realizations. So the question arises as to how these two results fit together. The answer is that for a given n, solutions derived from a given arrangement of reflection coefficient zeros correspond to those obtained from a specific section of the total range of d. By accounting for all the different arrangements of zeros, the range of the d axis from zero to $d_{max}$ is covered. For example, if all zeros are restricted to the left half plane, the resulting continuum of networks has values of d that are between zero and something less than $d_{max}$. Other zero arrangements cover other adjoining sections of the complete range of d.

For a particular transfer function, the zeros of the reflection coefficient $\rho(\lambda)$ change with the termination ratio $R_\ell/R_s$. The n zeros of $\rho(\lambda)$ are chosen from the 2n zeros of $\rho(\lambda)\,\rho(-\lambda)$, so by investigating the behavior of the real zeros of this latter function as the termination ratio changes we can determine the number of different networks possible. This has been done for Bessel transfer functions of two, three, four, and five poles, and the results are as follows.

Let $\Theta = R_\ell/R_s$.

(1) BE 2. For $1/3 \leqslant \Theta \leqslant 3$, $\rho(\lambda)\,\rho(-\lambda)$ has all (four) real zeros, and (except for $\Theta = 1$) two different networks are possible. For $\Theta = 1$, $\rho(\lambda)\,\rho(-\lambda)$ has a double zero at $\lambda = 0$ and a pair of simple zeros at $\lambda = \pm\sqrt{3}$, so only one network is possible. For $\Theta = 1/3$ or 3, $\rho(\lambda)\,\rho(-\lambda)$ has double zeros at $\lambda = \pm\sqrt{1.5}$. For $\Theta < 1/3$ or $>3$, the zeros are complex and again only one network is possible. By calculating from the data in Weinberg's tables for lowpass filters (Ref. 7, p. 618) for all zeros of $\rho(\lambda)$ in the left half plane, it is found that $0 \leqslant \Theta \leqslant 1$ and the corresponding range of d is $0 \leqslant d \leqslant 0.634$. For the remainder of the range of d, $0.634 \leqslant d \leqslant 1.5$, the data given in Fig. 51 shows that $1 \leqslant \Theta \leqslant 3$, which is the second solution mentioned above, apparently not observed by Weinberg.

(2) BE 3. For this transfer function, $\rho(\lambda)\,\rho(-\lambda)$ has two real and a quadruplet of complex zeros for all values of $\Theta$, so (except for $\Theta = 1$) there are two solutions. For $\Theta = 1$ the two real zeros occur as a double zero at $\lambda = 0$, thus only one solution is possible. Reference to Weinberg's tables shows that for zeros in the left half plane with $0 \leqslant \Theta \leqslant 1$, the range of d is $0 \leqslant d \leqslant 0.7968$, and for alternating zeros with $1 \leqslant \Theta \leqslant \infty$, the range of d is $d \geqslant 0.7968$. The upper limit on d cannot be found from Weinberg's tables, but the data presented in Figs. 76 through 79 show that it is 2.322. The figures show that for these two ranges of d, $\Theta$ is in agreement with Weinberg's result.

(3) BE 4. This case has more possibilities. If $0.1723 \leqslant \Theta \leqslant 5.802$, $\rho(\lambda)\,\rho(-\lambda)$ has four real and a quadruplet of complex zeros, so (except for the two special cases cited below) there are four solutions. If $\Theta = 1$, two of the real zeros coalesce into a double zero at $\lambda = 0$; thus only two solutions are possible. If $\Theta = 0.1723$ or 5.802, the four real zeros coalesce into two double zeros at $\lambda = \pm 2.544$, and only three solutions are possible. For $\Theta < 0.1723$ or $>5.802$, all zeros of $\rho(\lambda)\,\rho(-\lambda)$ are complex and two solutions are possible. Again by referring to Weinberg's tables we find that for zeros of $\rho(\lambda)$ in the left half plane and $0 \leqslant \Theta \leqslant 1$, the range of d is $0 \leqslant d \leqslant 0.9436$. For alternating zeros of $\rho(\lambda)$ and $0 \leqslant \Theta \leqslant 1$, the range of d is $4.24 \leqslant d \leqslant 5.76$. Obviously this second solution corresponds to the double solution given in Figs. 115 through 119 for $4.24 \leqslant d \leqslant 5.0$, and a few computations verify this. Several computations over the remainder

of the d-range show that as d goes from 0.9436 to 4.24, $\Theta$ varies from 1 to a maximum of 5.80 at d = 2.75 and then decreases to 1. This range, therefore, gives the remaining two solutions predicted above.

(4) BE 5. The situation in this case is relatively simple since $\rho(\lambda)\ \rho(-\lambda)$ always has exactly two real zeros and two sets of quadruplet zeros. For $\Theta = 1$, there is a double zero at $\lambda = 0$ and two solutions are possible. Otherwise, four solutions are possible. For all zeros of $\rho(\lambda)$ in the left half plane, Weinberg's tables show that for $0 \leqslant \Theta \leqslant 1$, the range of d is $0 \leqslant d \leqslant 1.075$. This range of d, shown in Figs. 162 through 167, gives $0 \leqslant \Theta \leqslant 1$, thus the two agree. For alternating zeros Weinberg's data leads to values of d that fall into two parts of the realizable d range (for a lossless filter the range $3.60 < d < 4.65$ is not realizable); for $\Theta = 8$, d = 3.08 and for $1/4 \leqslant \Theta \leqslant 1$, d is in the range $5.28 \leqslant d \leqslant 6.16$. These design data have the following regions. For $1.075 \leqslant d \leqslant 3.60$, the $\Theta$-range is $1 \leqslant \Theta \leqslant \infty$. For $4.65 \leqslant d \leqslant 7.50$, the $\Theta$ range is $0 \leqslant \Theta \leqslant 4.90$; and for $6.80 \leqslant d \leqslant 7.50$, the $\Theta$-range is $0 \leqslant \Theta \leqslant 0.204$. The latter range is actually a foldover of the previous range if it were extended above d = 7.50. All these data are consistent with Weinberg's, but his data are not extensive enough to permit a definition of the various regions, as was possible in the previous cases. Weinberg's set of data for $\Theta = 8$ is not consistent with the remainder of his data for alternating zeros, so possibly this one design does not actually possess alternating zeros.

Thus, for Bessel transfer functions it has been shown that certain ranges of d correspond to certain arrangements of the zeros of the reflection coefficient. Except for the few instances noted above, the correspondence is complete.

## REFERENCES


1. R. Feldtkeller, Einführung in die Theorie der Hochfrequenz-bandfilter, 4 Aufl. (S. Hirzel, Stuttgart, 1953).

2. E. Green, Amplitude Frequency Characteristics of Ladder Networks (Marconi's Wireless Telegraph Co., Ltd., Chelmsford, Essex, 1954).

3. Reference Data for Radio Engineers, 4th ed. (International Telephone and Telegraph Corporation, New York, 1956).

4. H. Meinke and F.W. Gundlach, Eds., Taschenbuch der Hochfrequenztechnik (Springer, Berlin, 1956), Ch. N, Secs. 7-12, by H. Behling.

5. M. Dishal, "Gaussian-Response Filter Design," Elec. Comm. 36, 3-26 (1959).

6. R. Saal, Der Entwurf von Filtern mit Hilfe des Kataloges normierter Tiefpässe (Telefunken G.M.B.H., Backnang/Württ., 1961).

7. L. Weinberg, Network Analysis and Synthesis (McGraw-Hill, New York, 1962).

8. A Handbook on Electrical Filters (White Electromagnetics, Inc., Rockville, Maryland, 1963).

9. P. Geffe, Simplified Modern Filter Design (Rider, New York, 1963).

10. G.L. Matthaei, L. Young, and E.M.T. Jones, Microwave Filters, Impedance-Matching Networks, and Coupling Structures (McGraw-Hill, New York, 1964).

11. G. Fritzsche and G. Buchholz, "Filterkatalog und Anwendungen des Filterkatalogs," Nachrichtentechnik 14 (April 1964) to 15 (February 1965), 15 (May-July 1965), and 15 (September-November 1965).

12. M. Dishal, "Design of Dissipative Band-Pass Filters Producing Desired Exact Amplitude-Frequency Characteristics," Proc. IRE 37, 1050-1069 (Sept. 1949).

13. T.C.G. Wagner, "The General Design of Triple- and Quadruple-Tuned Circuits," Proc. IRE 39, 279-285 (March 1951).

14. A. Papoulis, The Fourier Integral and Its Applications (McGraw-Hill, New York, 1962).

15. P.R. Geffe, "A Note on Predistortion," Trans. IRE, PGCT CT-6, 395 (Dec. 1959).

16. H.J. Orchard, "Predistortion of Singly-loaded Reactance Networks," Trans. IRE, PGCT CT-7, 181-182 (June 1960).

17. C.A. Desoer, "Notes Commenting on Darlington's Design Procedure for Networks Made of Uniformly Dissipative Coils ($d_o + \delta$) and Uniformly Dissipative Capacitors ($d_o - \delta$)," Trans. IRE, PGCT CT-6, 397-398 (December, 1959).

18. W. Cauer, Theorie der Linearen Wechselstromschaltungen, Band II (Akademie, Berlin, 1960), Chap. 20 by N.T. Ming.

19. V. Belevitch, J.M. Goethals, and J. Neirynck, "Darlington Filters with Unequal Uniform Dissipation in Inductors and Capacitors," Philips Res. Rep. 19, 441-468 (October, 1964).

20. M. Dishal, "Alignment and Adjustment of Synchronously Tuned Multiple-Resonant-Circuit Filters," Proc. IRE 39, 1448-1455 (Nov. 1951).

21. H. Takahasi, "On the Ladder-Type Filter Network with Chebyshev Response," J. Inst. Elec. Comm. Engrs. Japan 34 (Feb. 1951), in Japanese; English translation: L. Weinberg and P. Slepian, "Takahasi's Results on Tchebycheff and Butterworth Ladder Networks," Trans. IRE, PGCT CT-7, 88-101 (June 1960).

22. L. Weinberg, "Explicit Formulas for Tschebysheff and Butterworth Ladder Networks," J. Appl. Phys. 28, 1155-1160 (Oct. 1957); also IRE Natl. Conv. Record 5, Pt. 2 (1957).

INDEX OF DESIGN CHARTS

GENERAL INDEX